Learn Geometry NOW!

Geometry for the Person Who Has Never Understood Math!

Minute Help Guides

Minute Help Press

www.minutehelp.com

Table of Contents

Introduction

Geometry is that section of math that deals with lines and shapes instead of just numbers. If you have trouble understanding geometry, you're not alone! Plenty of people have had trouble with geometry, even those skilled at other branches of mathematics. So if you can't look at a triangle without flinching, don't worry—help is on the way. Throughout the next twelve sections, you will go over the major geometrical concepts, from points and lines to three-dimensional solids. Geometry isn't impossible; you *can* and *will* learn it. Just believe in yourself, and you might find that geometry isn't as difficult as you thought.

Chapter 1: Points, Lines, and Planes

Do you remember seeing one of these?

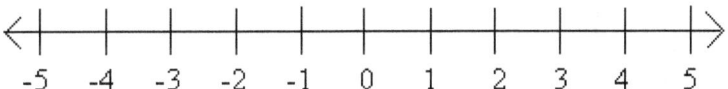

This is a *number line*, as you may already know. Just by looking at it, it's pretty obvious how it works: each vertical segment represents a number, and the numbers get bigger the farther right you go. Because there are an infinite amount of numbers, the number line also stretches into infinity. The number line is also *one-dimensional*: it goes only left and right.

The most basic concept in geometry, necessary to understanding all the other concepts, is quite similar to the old number line. It's called a "cartesian plane," and it looks like this:

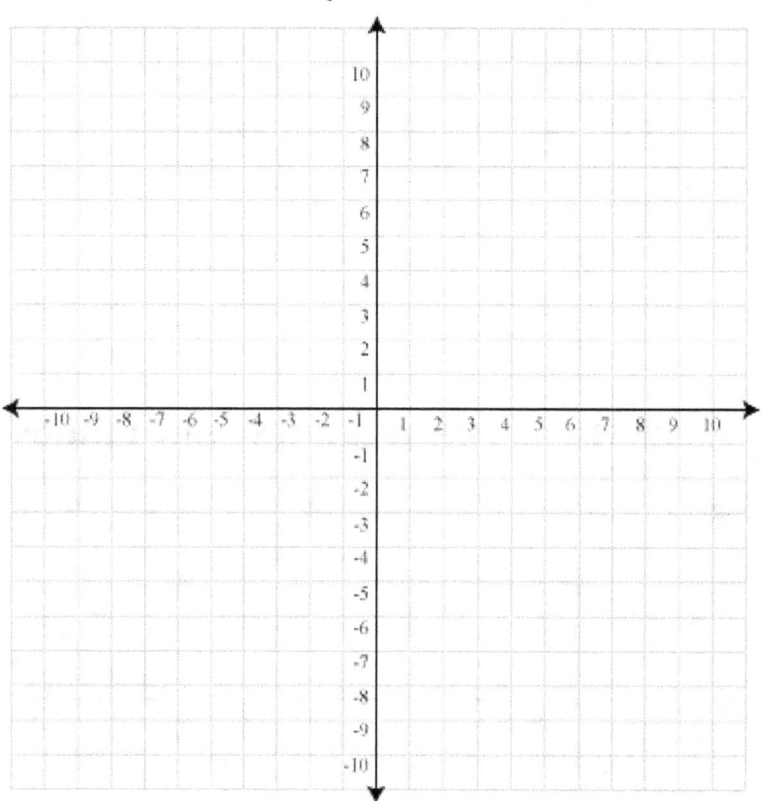

As you can see, the cartesian plane is really just two number lines slammed together. The horizontal line is exactly like a number line and is called the "x-axis." The new line, the vertical one, is called the "y-axis," and numbers on it get bigger the farther up you go. Unlike the number line, it is *two-dimensional*: it goes right, left, up, and down.

The cartesian plane will be used to explain every concept we discuss in these articles. First, we'll use it to discuss points, lines, and planes.

Points

A *point* is a zero-dimensional object with no extension in any direction. It's just a dot. It doesn't take up any space; it's just a location.

Consider the following: (1,2). This is an *ordered pair* of numbers: it's two numbers, set in an order that matters—that is, (1,2) is different from (2,1). We can represent (1,2) on a cartesian plane like this:

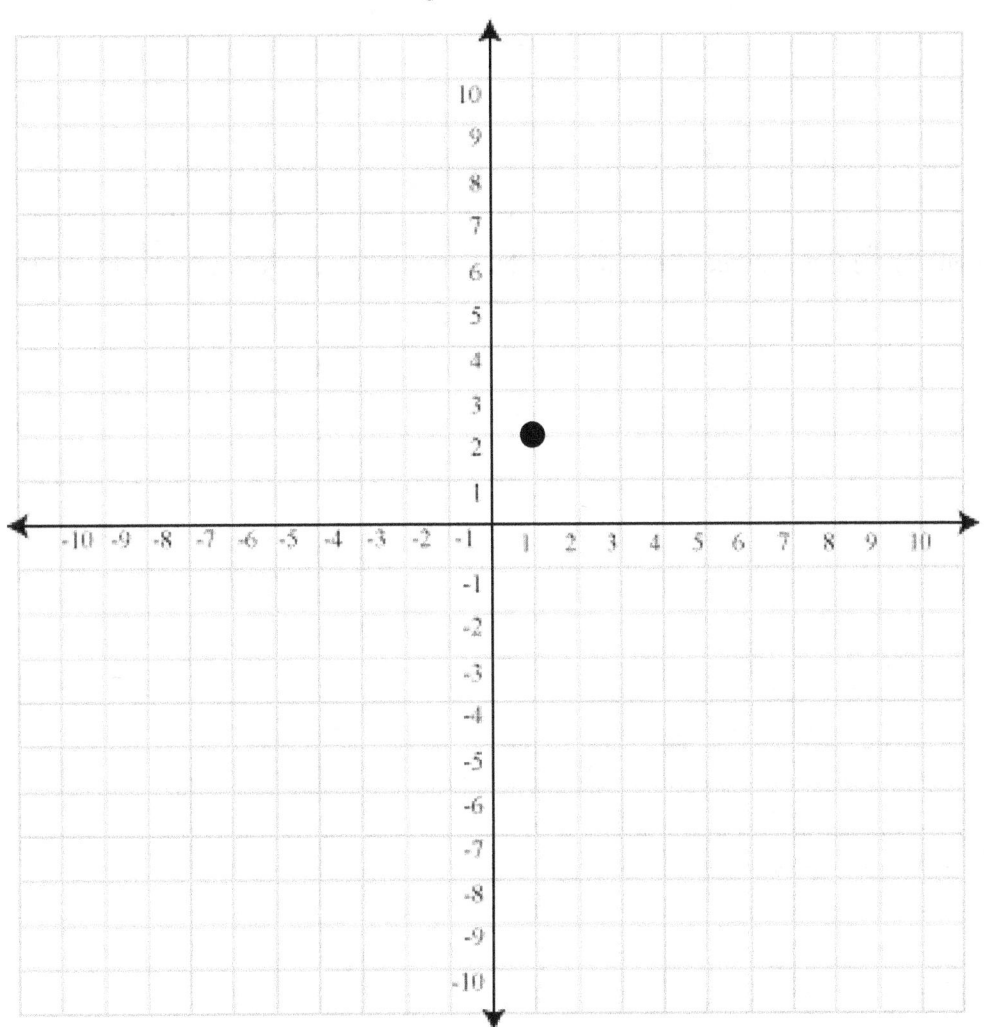

(1,2) refers to the x and y-axes, in that order—they're called the *x-coordinate* and the *y-coordinate*, respectively. We start at the *origin*—the point where the lines intersect, or (0,0). We then move *left* or *right* until we get to the first number of the ordered pair; following that, we move *up* or *down* until we get to the second number.

It should be noted that the dot on that graph up top is only an approximation and a rather crude one at that. An actual point, as stated before, is a zero-dimensional object: it does not stretch at all in space, in any direction. After all, the point (1,2) is different from the point (1.00001,2), or (1.0000000001,2), or even (1.<100 trillion zeroes)1,2). It is always going to be impossible to draw a point on a plane without that point bulging over onto the territory of nearby points. Just keep in mind that points are infinitely small, and you should be fine.

Lines

You've heard this expression: "The shortest distance between two points is a straight line." This is true in geometry as well. Indeed, when you say "line" in geometry, you always mean a straight line. Non-straight lines are called *curves*—but that's for later.

You can make one (and only one) line between any two points. For example, here's a line between (0,0) and (2,2):

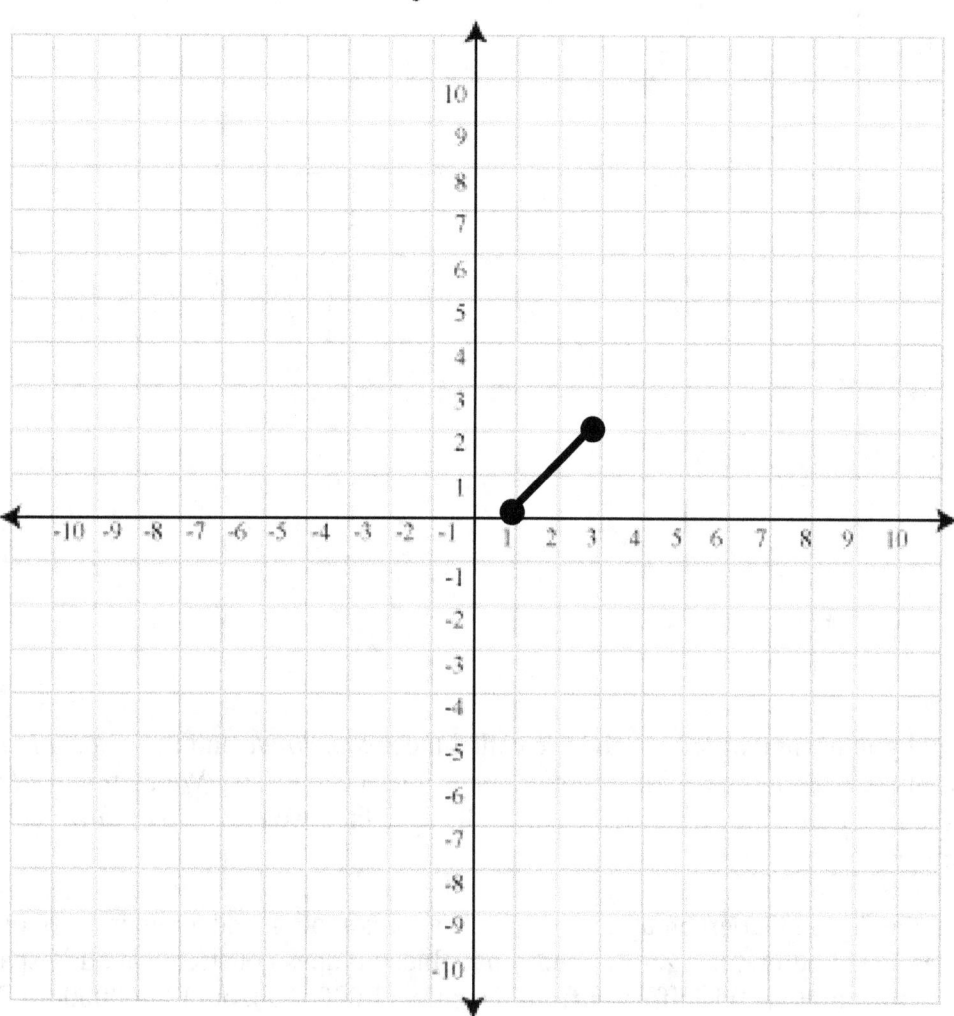

Here, (0,0) and (2,2) are called *endpoints*, as they're the points at the ends of the line.

Actually, though, a line with endpoints like this isn't usually called a line, it's called a *line segment*. A proper line, like the x and y-axes, goes on forever in each direction. For example:

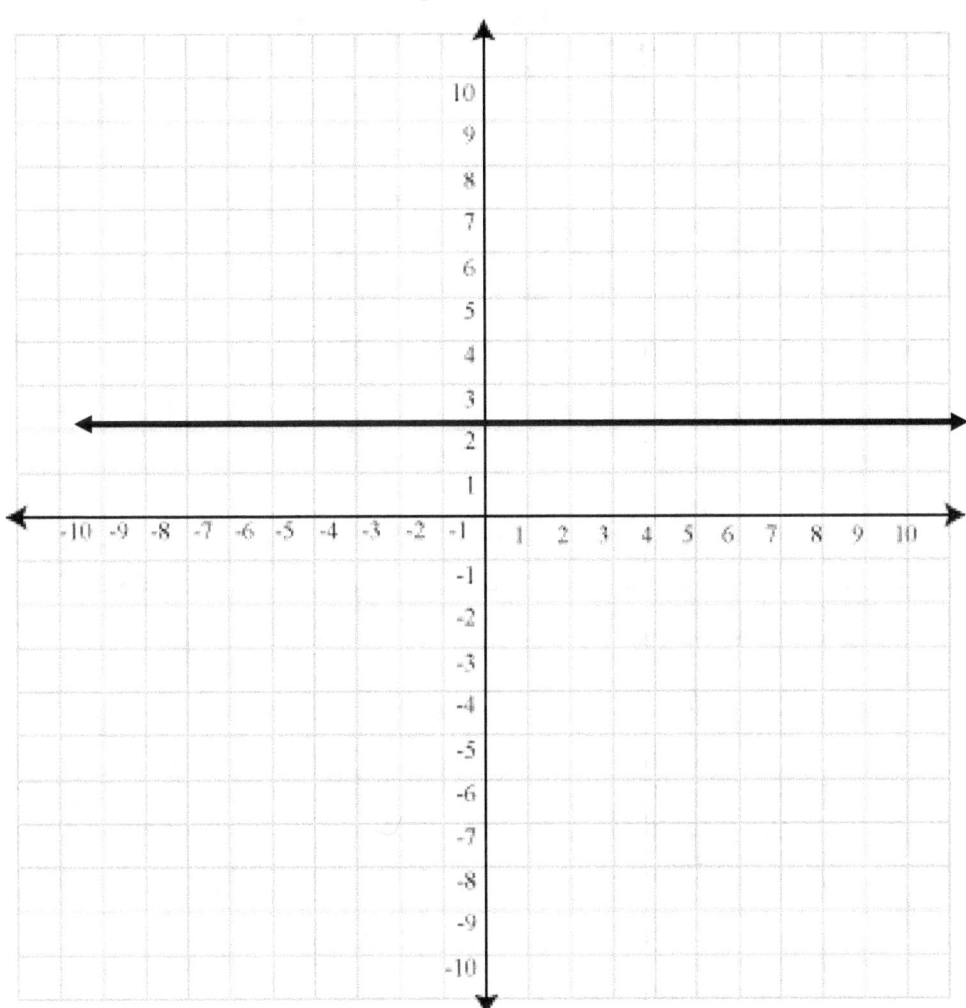

On the figure above, the arrows at each end signify that the line goes on into infinity into each direction.

So what exactly is a line, then? A line is a *one-dimensional object*, as it extends in two directions. Lines are made up of points—an infinite number of points, to be exact. (Since points are infinitely small, you need an infinite number of them to make something.)

Because lines have extensions, unlike points, they also have lengths. So a proper line has infinite length. Line segments, though, have finite lengths. How do you find the length?

It's easy to figure out some of them. For example, consider the line segment between (-1,3) and (-1,-4):

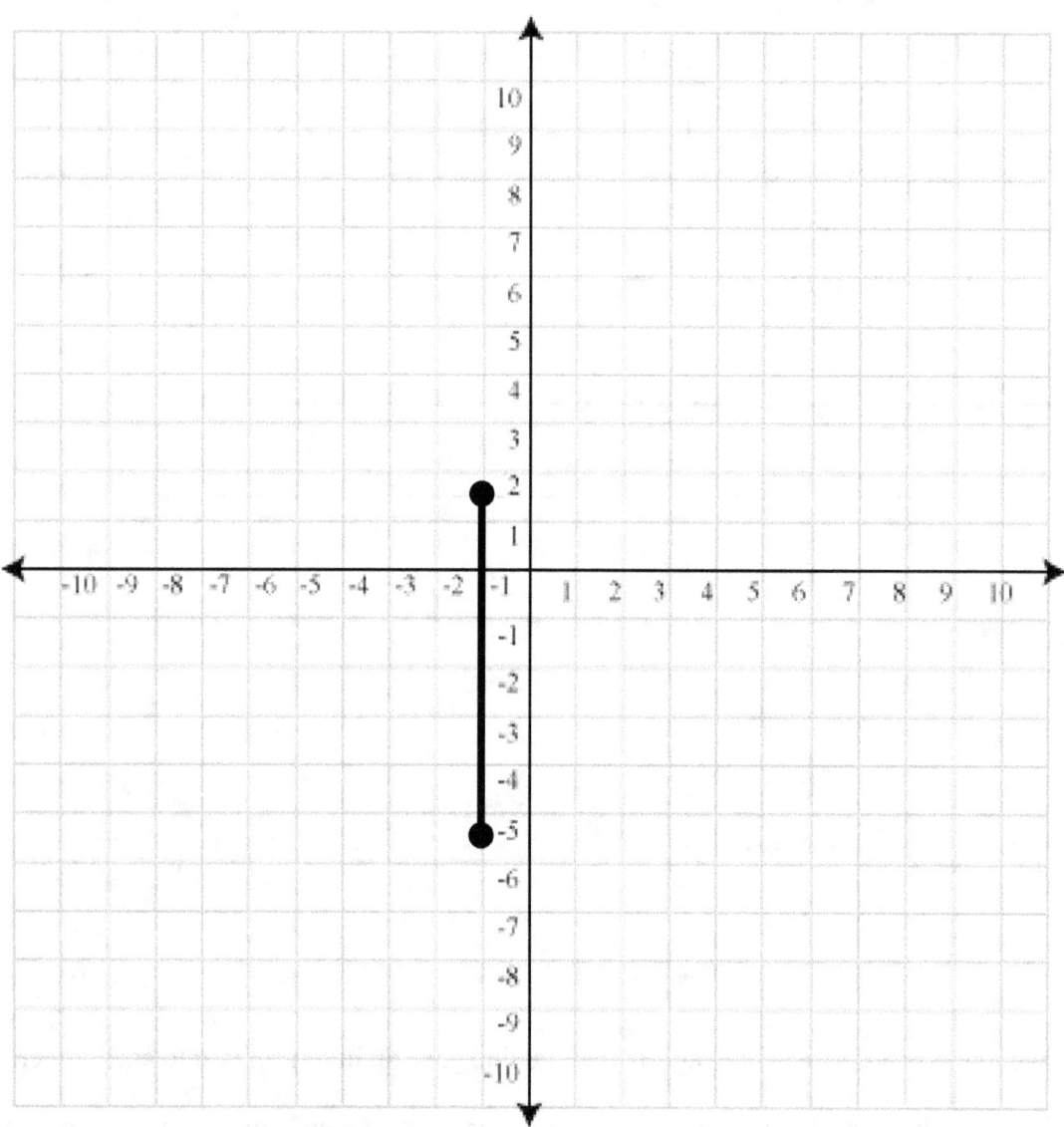

You can count for yourself and find that the length is 7.

What about a diagonal line segment, though, like the one between (1,2) and (3,5)? There is a way to find the length of this, too, but you'll have to wait for the section on right triangles to find out about it.

One final word about lines: Whenever two lines intersect, they always intersect at a point. For example:

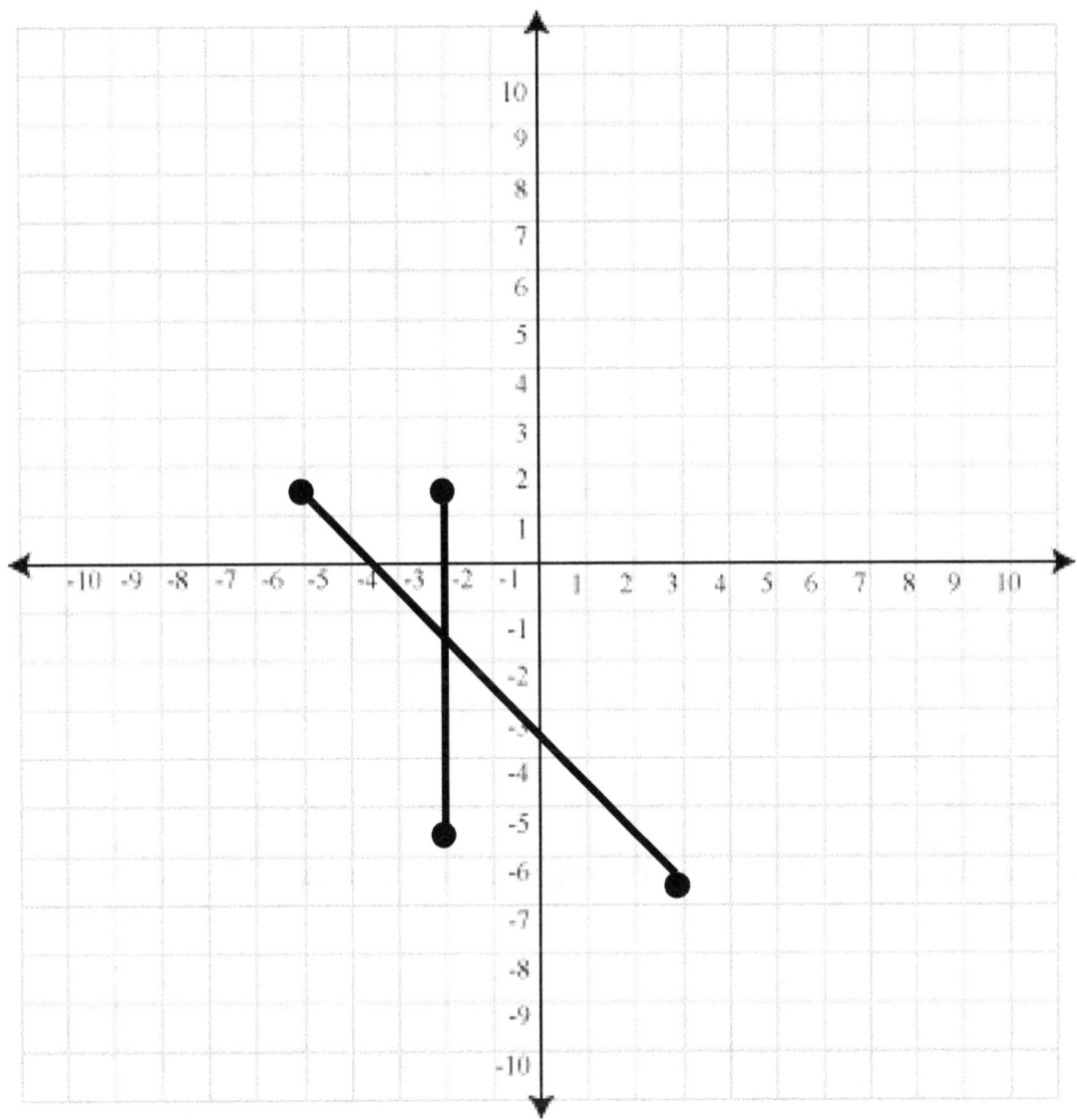

The two lines intersect at one, and *only* one, point: (-2, -1). For this reason, another definition of "point" is "the intersection between two lines."

Planes

The move from lines to planes mirrors the move from points to lines. Planes are two-dimensional objects that extend forever in all directions. You can think of a piece of paper or a wall as a plane—well, you could if they were infinitely large. But at least they *look* like planes.

A plane is made up of an infinite number of lines – remember that lines take up space only in two directions, not all four. A line is like an infinitely long piece of string.

Two planes intersect at a single line. Look at where two walls meet, and you will see a single line.

In most of geometry, the only plane you'll have to deal with is the cartesian plane itself. This is because most of geometry deals with two-dimensional objects like squares and circles, which can happily exist on a single plane. But when we do three-dimensional graphs at the end, planes will become very important.

Chapter 2: Deductive Reasoning

In geometry, one of the most basic ideas is the concept of a *proof*. A proof is a form of deductive reasoning, so that's what we'll talk about in this section.

Consider the following sentence: "The sun will rise tomorrow." You probably believe that sentence is true. But if someone asked why you believe it's true, what would you say?

What that person is really asking for is *justification* for your belief. If a belief has no justification—if there's no reason to believe it's true—well, we probably shouldn't believe in it, right? So it's very important for us to justify our beliefs.

Let's return to the original sentence. There are several ways to justify the belief that "the sun will rise tomorrow." "The sun has always risen in the past" is a good one. You could also tell some story about astronomy, how the Earth revolves around the sun and rotates and such.

But notice something about those justifications: They don't *guarantee* the truth of the statement "The sun will rise tomorrow." After all, just because something happens in the past doesn't mean it'll happen in the future. In a few hours, the sun could go out or the Earth could hurtle off its orbit. So while the justifications just offered might make "The sun will rise tomorrow" *likely*, they don't make it *certain* with 100% accuracy.

Arguments similar to how we defended "The sun will rise tomorrow" are examples of *inductive logic*. Inductive logic reasons from the past to argue about what will happen in the future. We use inductive logic a lot (including in all of science!), but it can't justify anything completely.

Deductive logic is different. Consider the following argument:

"The sun is a star. Therefore, the sun is a star."

Deductive logic, unlike inductive logic, gives certain justification. If it is true that the sun is a star, then it is true that the sun is a star, and we know that with 100% accuracy. Deductive logic reasons from the *premises* (the things that the argument assumes to be true) to the *conclusion* (the thing the argument is trying to justify)—and it does this according to strict logical rules that *always* hold.

Now, consider the following:

"2 + 2 = 4"

Is this an example of deductive or inductive logic?

Deductive logic, of course. When we assert that $2 + 2 = 4$, we're not saying that since $2 + 2$ has equaled 4 in the past, it'll equal 4 now. We're saying $2 + 2$ *necessarily* equals 4—that whenever you add 2 and 2, from the beginning of time to its end, you'll always get 4. It's a classic example of deductive logic.

Unfortunately, the things you'll have to prove in geometry aren't as easy as what 2 and 2 equal! Here's a slightly more complex example, in the format that you might receive when asking for proofs in geometry:

The line segment L1 has endpoints (2,1) and (2,6). The line segment L2 has endpoints (-3,2) and (-3,-5). Prove that the length of L1 plus the length of L2 equals 12.

Remember: In order to prove something deductively, the argument we make has to hold 100% of the time. It's not enough to just make the conclusion likely or plausible; it has to be *certain*. So what would a good answer to the above problem look like?

First, to make things easier for ourselves, let's draw L1 and L2 on a graph.

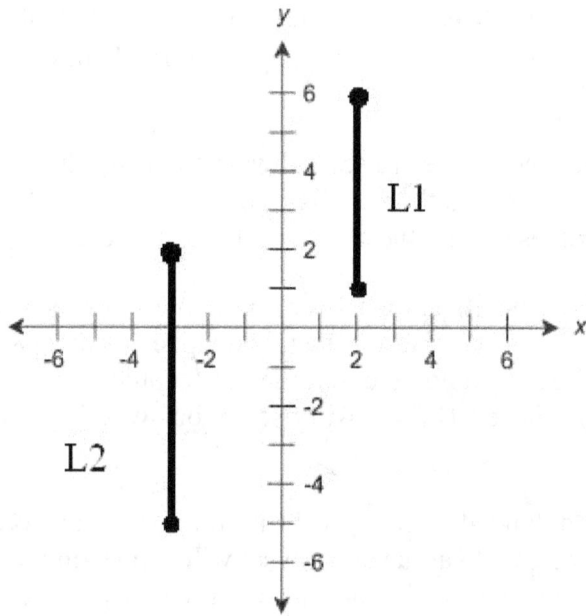

It should be fairly easy to see for yourself that the length of L1 is 5 and the length of L2 is 7. With that in hand, the proof becomes easy.

So how do you write out a proof? You start by stating the assumptions, or the premises, that were given to you by the problem, like this:

- The endpoints of L1 are (2,1) and (2,6). [Assumption]

- The endpoints of L2 are (3,2) and (3,-5). [Assumption]

Then, whenever you deduce something, you always list which line you got that deduction from:

- The length of L1 is 5. [from 1]

- The length of L2 is 7. [from 2]

If your deduction depends on two or more previous lines, just list them both:

- The length of L1 plus the length of L2 is 12. [from 3 and 4]

And then, now that you've arrived at the conclusion, you're done! Write this:

Q.E.D.

"Q.E.D." is an abbreviation of the Latin phrase "quod erat demonstrandum," which means "that which was to be demonstrated"; in other words, it's used to signal that you've arrived at the assertion you were supposed to demonstrate, or prove.

So that's all there is to it! Seems simple, right? Well, *that* proof may have been simple, but some mathematical proofs are so complicated that they take thousands of steps before finishing. Even if you don't want to be a mathematician, you'll probably be required to do a proof—or something like a proof, even if it's not called that—with 20 or more lines in it. But as long as you're careful, you should be fine. Onward!

Chapter 3: Parallel Lines and Planes

Parallel Lines

When talking about lines, we mentioned that two lines intersect at a point. But is it possible for two lines not to intersect?

Indeed it is! Consider the following:

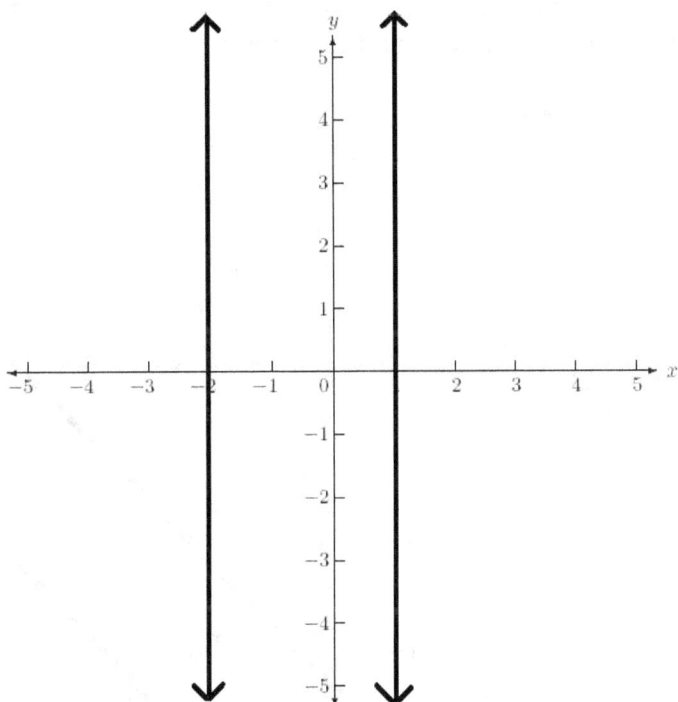

The line on the left contains all and only points whose x-coordinate is -2, and the line on the right contains all and only points whose x-coordinate is 1. In order for two lines to intersect, they have to *share a point*: there has to be a point that is the exact same in each of the lines. But according to the rules we set out, the lines can *never* have a point with the same x-coordinate and so can never have a point in common.

This gives us a definition of parallel lines: two lines are *parallel* if and only if they do not share a point.

(As a side note, you might wonder whether it's possible for two lines to share more than one point. The answer is *no*; two lines can intersect only once.)

So now that we have a working concept of parallel lines, let's look at some properties they have.

First of all, if we want to write shorthand that two lines are parallel, we use this symbol: $\|$

For example, "line AB is parallel to line CD" can be written as:

$$\overleftrightarrow{AB} \parallel \overleftrightarrow{CD}$$

(Note: "AB" and "CD" here refer to any two points on the line. This works because of the principle cited before, that lines can only intersect once: any two points designate a *unique* line.)

Second, perhaps the most important property of parallel lines is that they are always the same distance away from each other—they never get closer to or father away from the other. If they got closer, they'd eventually intersect. If they got farther away, then they must have intersected previously (remember that lines are completely straight).

This means that it's possible to designate one unique distance between any two parallel lines. If they are both completely horizontal or both completely vertical, you can see that the two lines in the first graph of this section are 3 apart from each other. But what if the lines are diagonal?

For example, consider the following two lines:

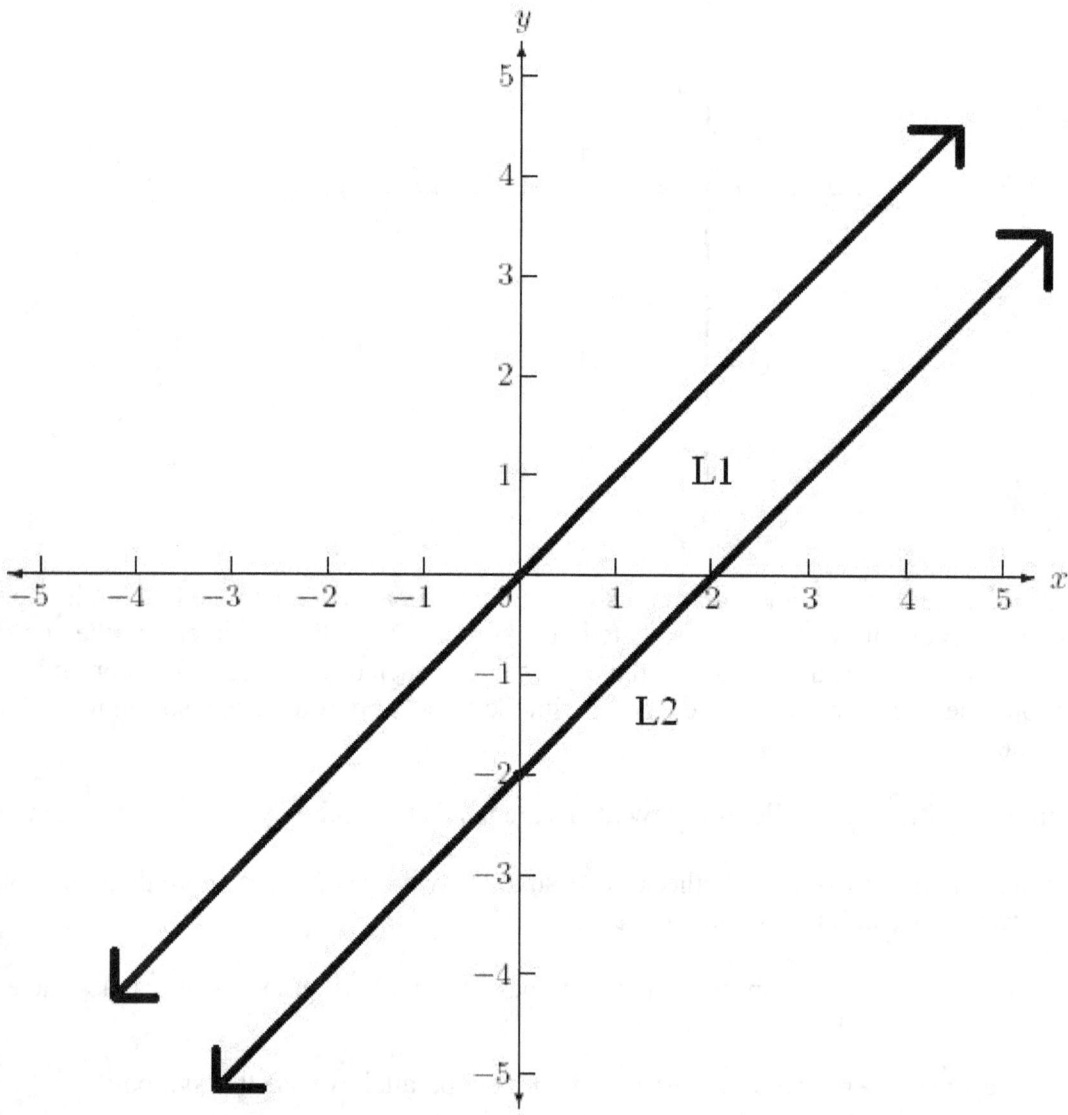

Every point in L1 has the same x-coordinate as its y-coordinate, and every point in L2 has a y-coordinate 2 less than its x-coordinate. It's easy enough to see that they're parallel, but what's the distance between them?

If you look at the axes between them, it might be tempting to say "2," but don't move so fast. In order to calculate the distance between L1 and L2, we first need to find a line that intersects them *at right angles*, or angles that look like this:

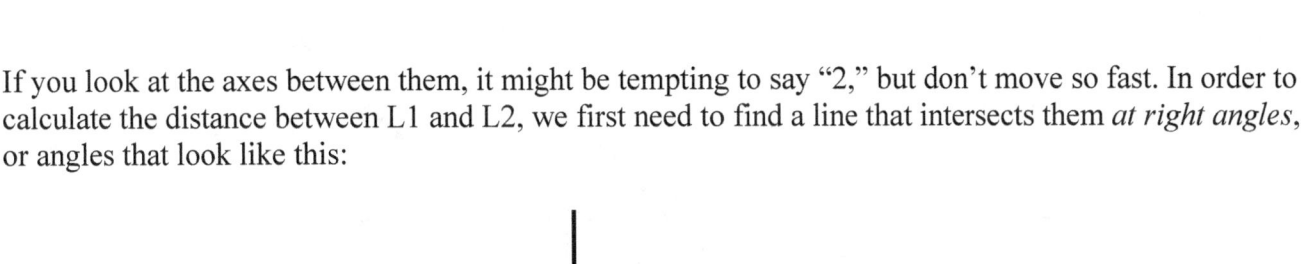

The space between the lines is called an angle, and when they intersect like this (you might notice that the axes on the cartesian plane intersect the same way), they're called "right angles." We'll have plenty more to say about these angles later.

As to the current problem, here's a good candidate:

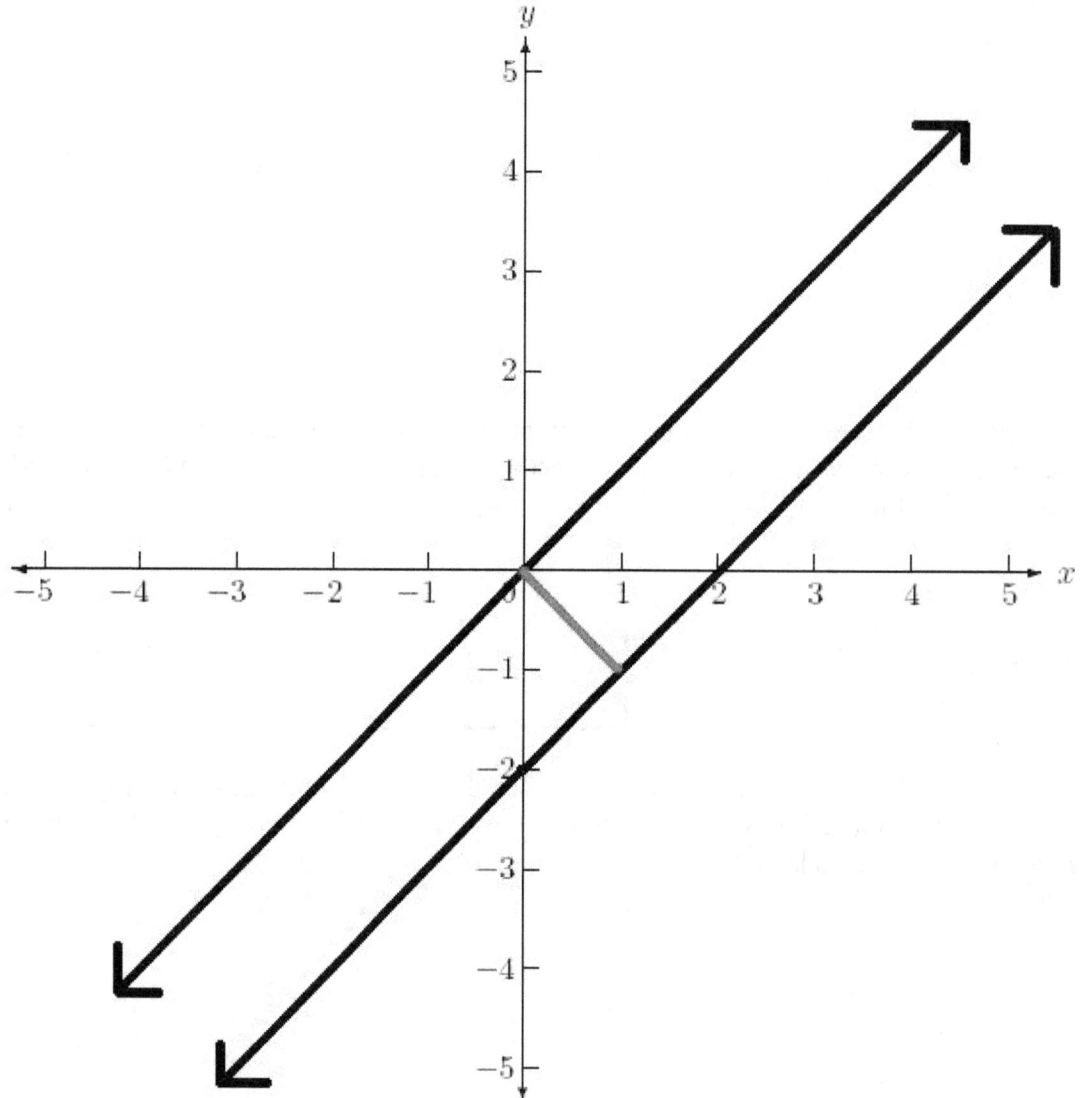

The length of the red line segment is the distance between the parallel lines. You might notice that the red line segment is a diagonal line segment, and I said earlier we need to learn about right triangles before being able to figure out their length. So this problem we must leave for now and return to in a later section, *Right Triangles*.

However, going through this process wasn't completely useless. Do you notice that the red line segment above intersects *both* L1 and L2 at a right angle? This isn't just a coincidence; it's another property of parallel lines. To generalize, consider the following diagram:

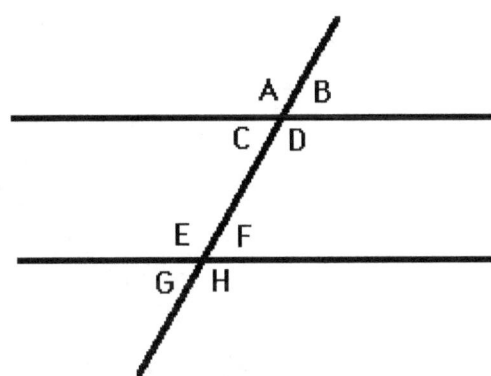

Assuming the two horizontal lines are parallel, then the following angles are equal to each other:

A, D, E, H *and* B, C, F, G. In other words, A = D = E = H and B = C = F = G.

Parallel Planes

One thing should be noted about parallel lines, however: two lines can be parallel *only* if they exist on the same plane. As mentioned previously, we'll almost exclusively be talking about the cartesian plane, so in these articles you can mostly ignore that criterion—but not always.

This might lead you to ask the question, can two planes (or a line and a plane) be parallel as well? The answer is yes, under the same conditions as lines: two planes (or a line and a plane) are parallel if and only if they do not share a single point. You can imagine opposite walls in a house, for example, as being parallel.

One last thing to note: this concept of lines and planes being "parallel," and what exactly that means, is key to geometry. Learn the tools of this lesson well.

Chapter 4: Congruent Triangles

Now that you know all about lines and planes, let's talk about geometrical objects. The first object we'll talk about is a three-sided object, or a *triangle*.

You probably already know what a triangle is. They look like this:

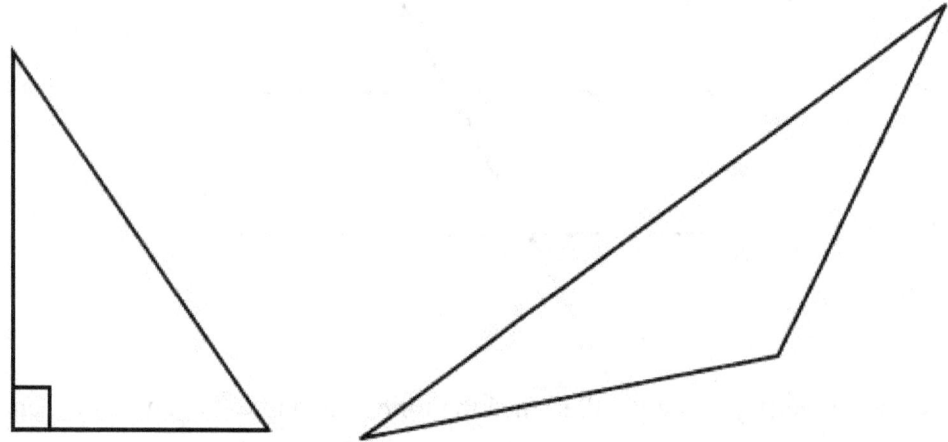

As you can see, triangles have three sides. Actually, if you take any three points and draw lines between them, you get a triangle. Those points are called *vertices*. On the triangles above, each place where two of the lines meet is called a *vertex*.

Triangles have various properties that you need to know. The most important has to do with angles, so let's take a small detour and talk about angles a bit.

You can think of an angle as the measurement of the way two lines intersect. So each vertex of a triangle has a corresponding angle. Just as we can measure the distance of lines, we can measure the size of angles in a unit called *degrees*. The sign for a degree is this: ° It's the same sign that you use for temperature.

To understand degrees, let's start with this angle.

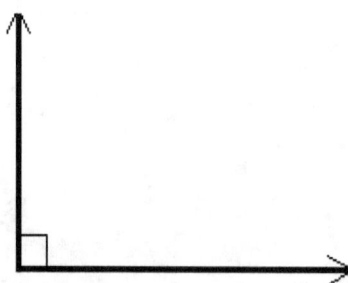

That's a 90° angle. This is also called a *right angle.* You can usually tell this, because people usually put a square between the lines instead of a circle, like above. (Triangles that have a right angle are called "right triangles," and you'll find out more about them later.) Cut a right angle in half, and you get a 45° angle; divide it into 9 parts, and each part will be 10°; and so on.

Now think about the angle created by the two lines below:

If an angle is just the space between two intersecting lines, then these two lines must have an angle as well. Indeed they do—it's the space above them. To make this easier to visualize, you can draw a semicircle above the line like this:

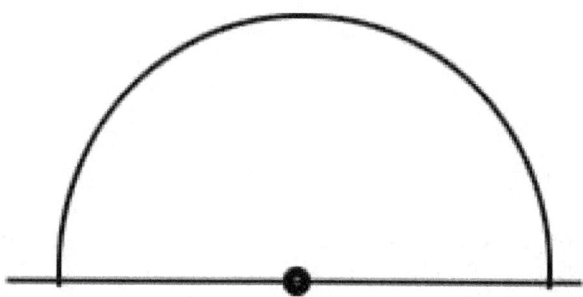

This angle, the space between two lines that intersect in this way, is *180 degrees* or *180°*. It is twice the measure of the 90° shown earlier.

Let's return to triangles. As you can tell, each triangle has three angles. In addition, in *every* triangle, the measurement of the three angles *must* add up to 180°. For example, if you know that two of a triangle's angles are 37° and 81°, the final angle must be 62°, as you see here:

$$37 + 81 + x = 180$$
$$118 + x = 180$$
$$118 - 118 + x = 180 - 118$$
$$x = 62$$

Now that you've covered all that material, you will be able to understand the title of this section: congruent triangles.

The basic way to think about congruent triangles is that they're the same. They don't have to take up the exact same positions on a cartesian plane, but you can *make* them exactly overlap just by doing one or more of the following things: moving, rotating, and/or flipping one of the triangles. Here's an example:

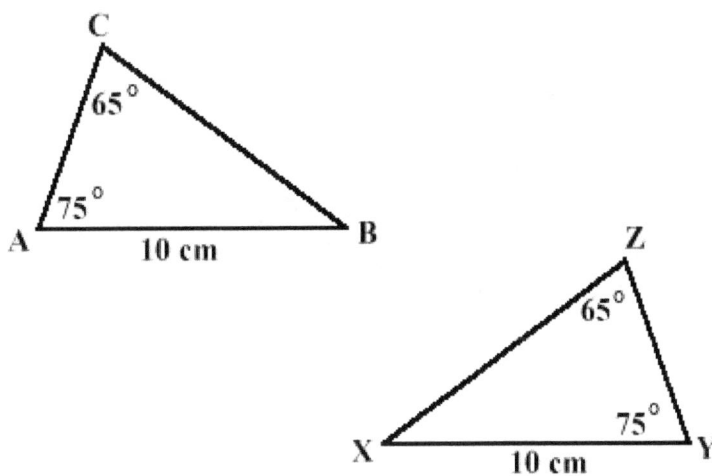

Assume the two triangles above are congruent (they at least look congruent!). Then, if you flip triangle XYZ horizontally and move it up and to the left, you can get it to precisely overlap triangle ABC. In that sense, they're the "same" triangle.

Now that you've got an intuitive idea of what congruent triangles are, let's give a mathematical definition of them:

Triangles are *congruent* if and only if their corresponding sides and angles are the same.

What does "corresponding" mean in this context? It means that each side and angle on one triangle has to "match up" with one—and exactly one—side or angle on the other in the same way.

Take the two triangles above as an example again. The line segments AB and XY are the same length: 10 centimeters. Let's assume, further, that the line segment AC is 6 centimeters. Since the angle on vertex A clearly matches up with the angle on vertex Y (they're both 75 degrees), side YZ *must* be 6 centimeters for the two triangles to be congruent. If side XZ is 6 centimeters instead, the triangles won't be congruent—the angles have to be between the *right sides*, as well as just being the same.

Now let's turn to one final topic: How can we tell when two (or more) triangles are congruent? Obviously, if we know that all six corresponding sides and angles are the same, we know the triangles are congruent by definition. But can we determine they're congruent with less information?

The answer is yes! Here are other ways to determine if triangles are congruent:

1. All three corresponding sides are equal. (SSS or side-side-side)

2. Two corresponding sides and the angle between them are equal. (SAS or side-angle-side)

3. Two corresponding angles and the side between them are equal. (ASA or angle-side-angle)

4. Two corresponding angles and a side that's *not* between them are equal. (AAS or angle-angle-side)

And if you're dealing with right triangles:

5. The side across from the right angle and one additional side are equal. (RHS or right-angle-hypotenuse-side)

These can be very useful. For example, if you know the corresponding sides are equal, then you also know that the corresponding angles are equal (and vice-versa).

So now you know much more about triangles!

Chapter 5: Quadrilaterals

We just learned about triangles, the only three-sided shapes—that is, the only three-sided *polygons*. Now let's talk about four-sided polygons.

Four-sided polygons are called *quadrilaterals*, as long as the sides are straight. The following are all quadrilaterals:

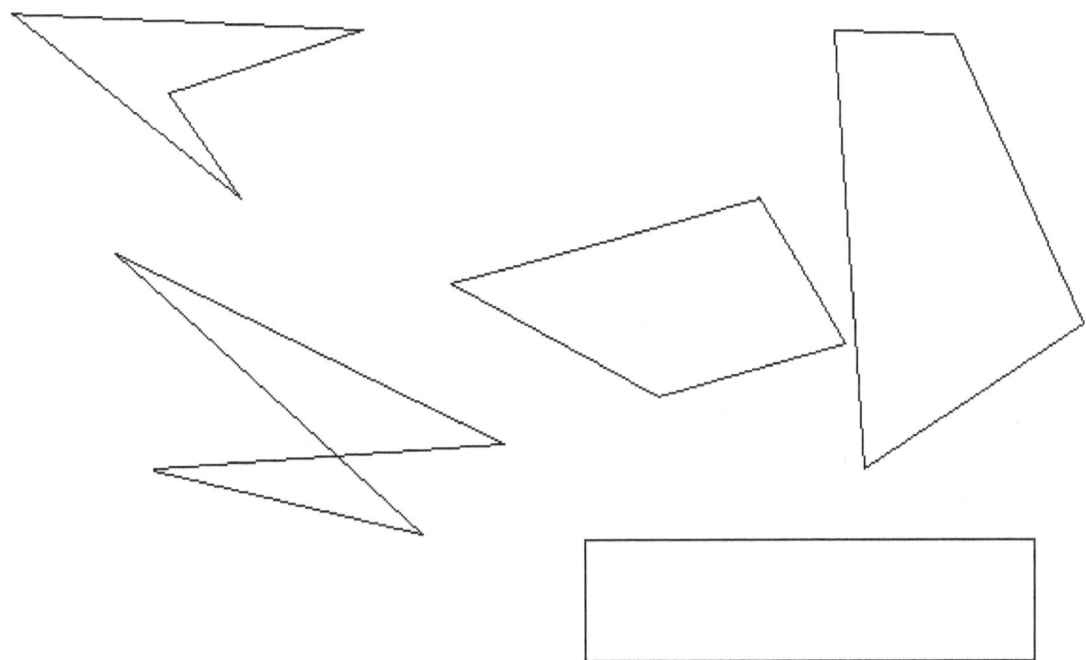

There is one property that all quadrilaterals share: *their four angles add up to 360 degrees*. However, while that's the only property common to all quadrilaterals, there are special types of quadrilaterals that possess additional unique characteristics. We'll talk about them in a bit, but first, let's talk about perimeters and areas.

The *perimeter* of a polygon is simply the length of all its sides. In other words, add up all the sides of a quadrilateral, and you'll get its perimeter.

The *area* is the space *inside* a polygon. The way you calculate the area depends on the polygon in question, so we'll talk about that for each individual unique quadrilateral.

Chapter 6: Parallelograms

A parallelogram is one of the most basic types of quadrilateral you need to know. As the name implies, the defining characteristic of a parallelogram is that its *opposite sides are parallel.* In addition, the opposite sides and opposite angles are *equal.*

In order to find the area of a parallelogram, you must first find its height. You can do this by drawing a line from one vertex (or angle) to the opposite side—making sure the line intersects the opposite side *at a right angle.* For example:

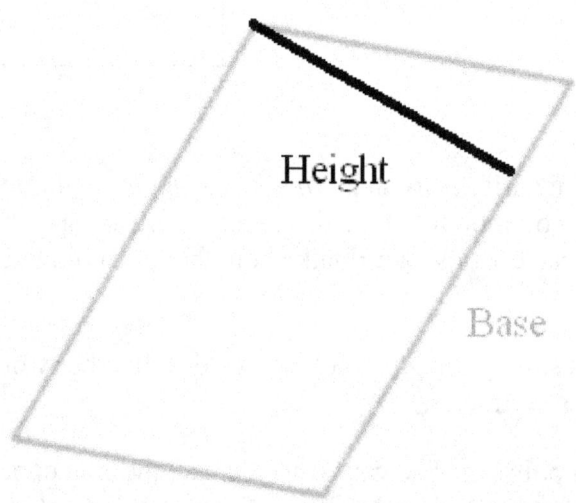

Then you multiply the base times the height. So if the height is 4 and the base is 7, the area is 4 x 7 = 28.

Chapter 7: Rhombus

The next three special quadrilaterals we'll cover are all parallelograms with additional limitations placed on them. We'll start with the rhombus.

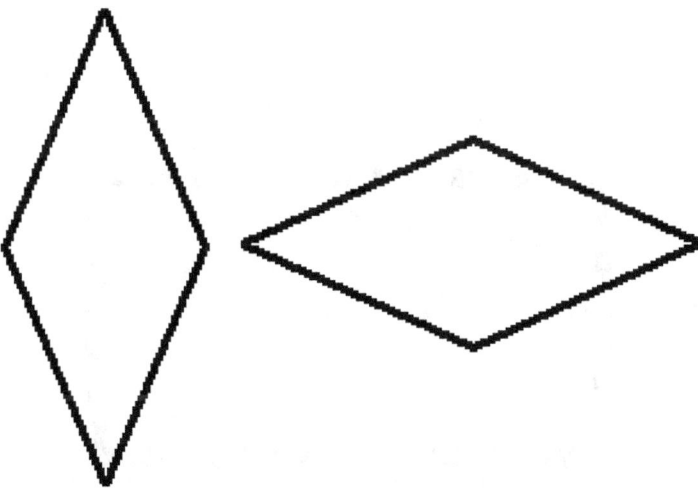

The rhombus is basically a parallelogram where *all four sides are equal*. Other than that, all its properties are the same as a parallelogram.

As a type of parallelogram, you can find the area of a rhombus by multiplying its base times its height. But since all four sides of a rhombus are equal, you can just multiply the height (shown as the red line in the below example) by the length of any of the sides.

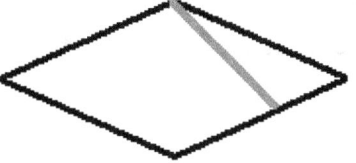

Chapter 8: Rectangles

You've probably heard of a rectangle before. A rectangle is basically a parallelogram where *all the angles are right angles* (90°). Since it's a parallelogram, opposite sides are also equal in length and parallel.

Because all of a rectangle's angles are right angles (90°), any side can function as a height. Therefore, to find a rectangle's area, just multiply the lengths of two sides that *aren't* parallel to each other.

Chapter 9: Squares

Squares are probably the easiest geometrical objects to deal with. A square can be thought of as:

A rectangle where all four sides are equal

A rhombus where all the angles are 90°

As such, all squares are also both rectangles and rhombuses.

Finding the area of a square is just like finding the area of a rectangle, except that all four sides are the same length. Since all sides are of equal length, multiply the length of one side by itself. For example, if each side is of length 8, the area of the square is 8 x 8 = 64.

Chapter 10: Trapezoids

(Note that in the UK, trapezoids are called "trapeziums." Here we will use the American English term.)

Other than parallelograms, there are two major types of quadrilaterals you should know about. The first is the trapezoid.

A trapezoid has only one qualification: At least one pair of opposite sides must be parallel. Any quadrilateral with one pair of sides parallel to each other counts as a trapezoid.

All parallelograms meet the above qualification, so all parallelograms are also trapezoids. Just keep in mind that *some* trapezoids are *not* parallelograms (like the one above).

Here's how to find the area of a trapezoid:

- Add the parallel sides and divide by two. In other words, find the average of the parallel sides.

- Then multiply the result by the height.

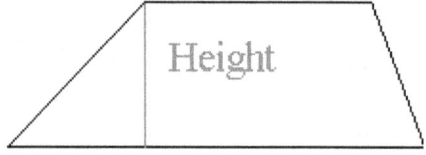

There is also a special type of trapezoid called an *isosceles trapezoid,* as shown here.

An isosceles trapezoid is a trapezoid where the non-parallel sides are equal in length. In addition, the pair of angles on each *parallel* side are equal. All rectangles are isosceles trapezoids—though *not* all *parallelograms* are.

Chapter 11: Kites

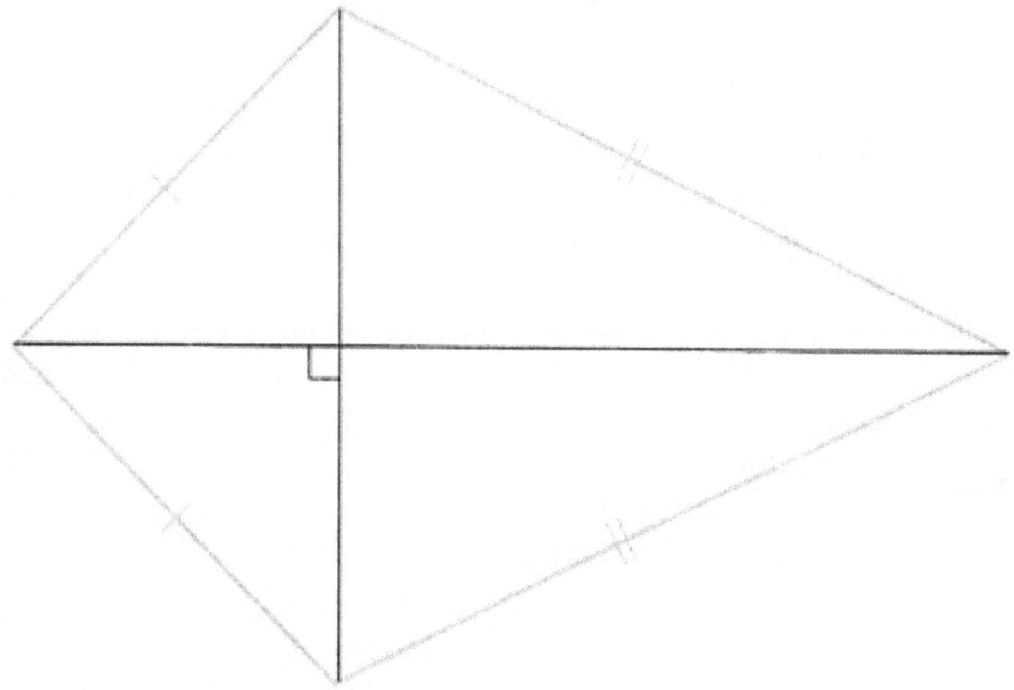

For a kite, *adjacent* sides and *opposite* angles have to be equal. In addition, the diagonals (lines joining opposite vertices, like the black lines above) have to intersect at right angles.

All *rhombuses*—including squares—are kites.

To find the area of a kite, multiply the lengths of the diagonals and divide by 2.

Summary

To help remember the relationships between quadrilaterals, let's draw a few Venn diagrams. Here's how they work: if Oval A (say the square oval) is inside Oval B (say the rectangle oval), everything that's an example of Oval A is also an example of Oval B (say, all squares are rectangles).

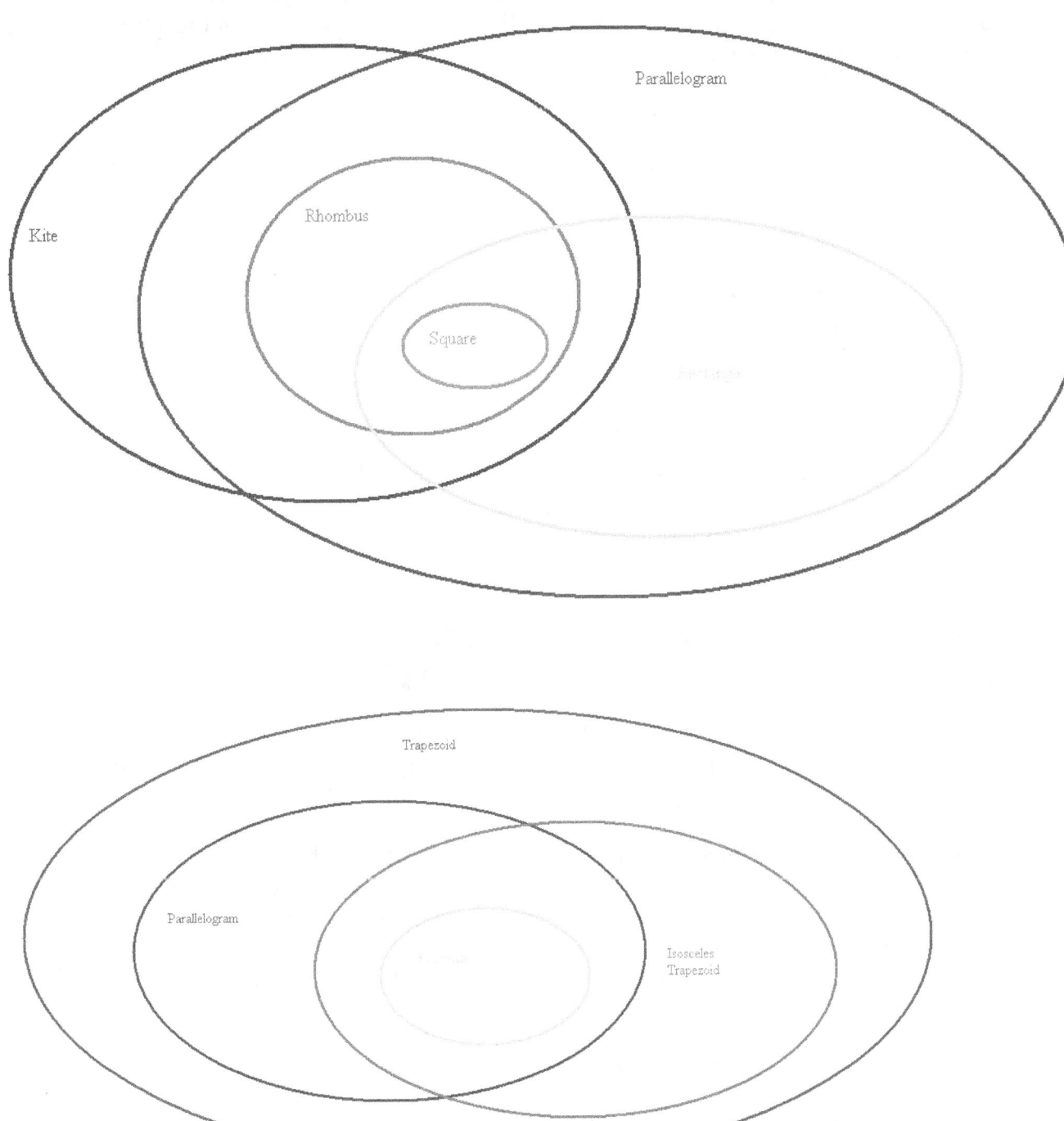

Parallelogram

Kite

Rhombus

Square

Rectangle

Trapezoid

Parallelogram

Isosceles
Trapezoid

Chapter 12: Inequalities in Geometry

If two numbers aren't equal to each other, their relation is one of inequality, right? But there are some inequalities in geometry that may not be obvious at first glance, mostly involving triangles. You'll find out more about them in this section.

Take a look at a typical triangle again.

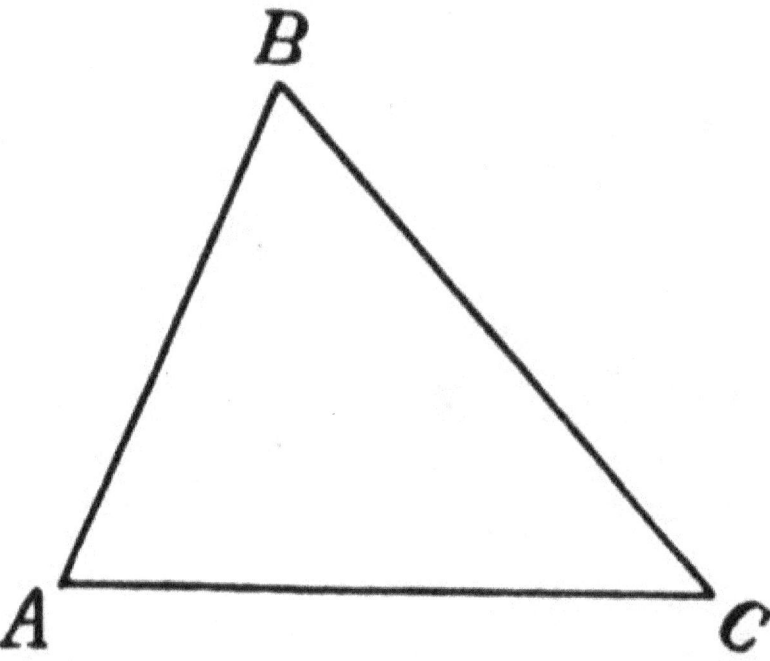

Consider the following: Would it be possible for one of the sides to be longer than *both* of the other sides put together?

The answer to this is *no*. The length of one side of a triangle is *always* shorter than the lengths of the other two sides added together. To see intuitively why this is, consider the following cliché: "The shortest distance between two points is a straight line." As it turns out, in geometry, this is true. So the shortest distance between A and B above is the line connecting them—the side AB. If you go across the other two sides to B, then, the distance you travel *must* be longer than the distance from A to B directly.

Note one consequence of this fact: a triangle with sides lengths 2, 2, and 4 cannot exist, because the longest side, 4, *equals* (and so isn't shorter than) the other two sides added together.

This inequality, often called the *Triangle Inequality Theorem*, has more uses than you might think, and you'll likely see it crop up on tests and the like. Just remember that the combination of two sides of a triangle is always greater than the third side, and you'll be fine.

Now, let's talk about how inequalities among the lengths of a triangle's sides affect that triangle's angles. Consider:

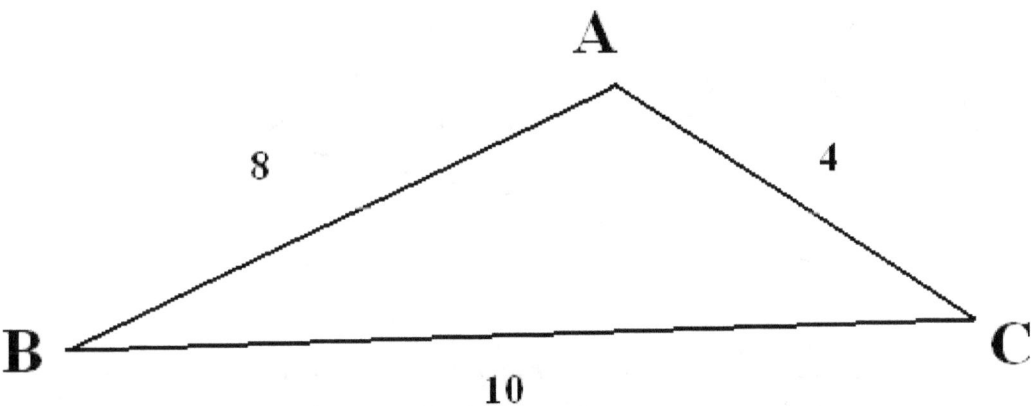

As you can see, side BC is greater than side AB, which is greater than side AC. Can we say anything about the relations between the angles?

Yes, we can! The largest angle of a triangle is the angle *opposite* the longest side (and vice versa—the longest side is opposite the largest angle). As you might expect, the middle angle is opposite the middle side (and vice versa), and the smallest opposite the shortest side (and vice versa).

So in the above triangle, the angle A (opposite side BC) is bigger than C (opposite side AB), which is bigger than B (opposite side AC).

Now then, for the last example of a geometric inequality we'll look at here, you'll need to learn about external angles. Consider this polygon:

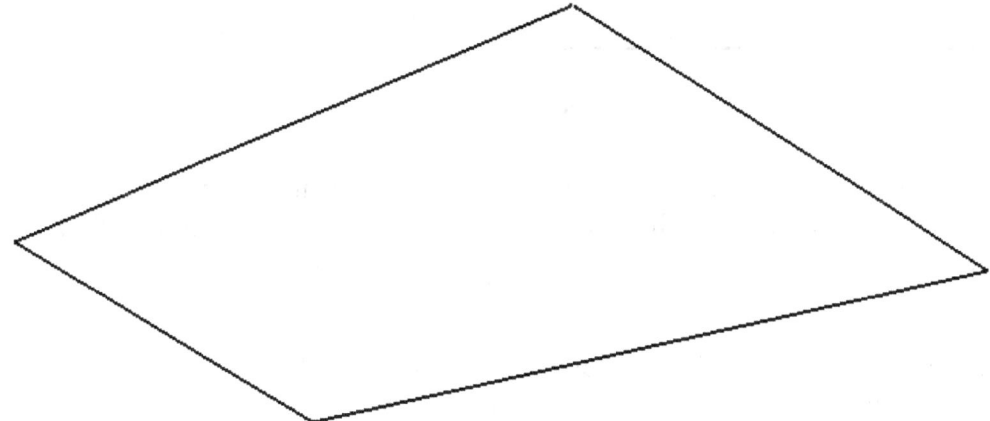

Now look at the vertex to the far right. To draw its *external angle*, just extend one of the sides that intersects at the vertex. For example:

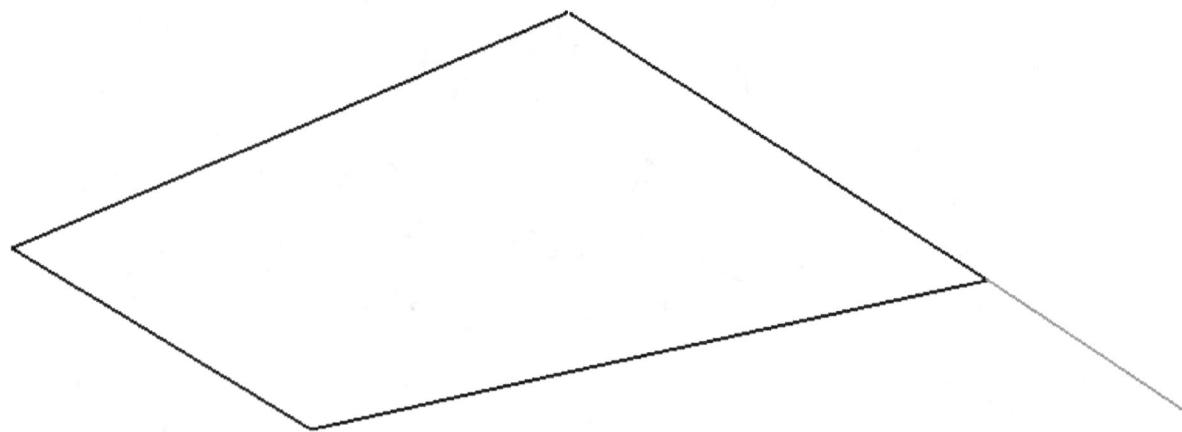

The new angle that's adjacent to the vertex is called an *external angle*. (Note that you can draw an external angle from either side that intersects with the vertex—each angle that results will be equal, for reasons that should become clear below.)

Notice something: the angle at the vertex and the external angle add up to the angle of a straight line. As we discussed before, the angle formed by a straight line is 180°. (Incidentally, two angles that form the angle above a line like this are called *supplementary angles*.)

Now let's try drawing an external angle for a triangle.

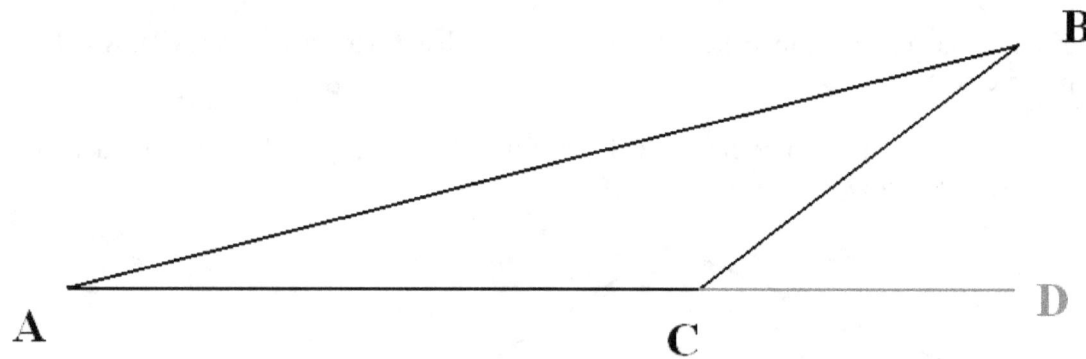

(Quick note: In order to pick out precisely which angle we're referring to in situations like the above, we can't just use the letters of the vertices like "C." Instead, we pick out an angle using a series of three letters, such as ACB. Read ACB as "the angle formed by AC and CB," just as BCD is "the angle formed by BC and CD." The middle letter is always at the vertex.)

Consider the external angle BCD and one of the other angles of the triangle, BAC. Question: Without any further information, can you say whether or not BCD is greater than BAC?

As it turns out, you *can*. This might seem not obvious, but think of this: You learned previously that the angles of a triangle add up to 180°. You also just learned that an external angle, plus the vertex it's adjacent to, add up to 180°. In other words, using the above triangle as an example:

ACB + ABC + BAC = 180

ACB + BCD = 180

Therefore

$$ACB + (ABC + BAC) = \qquad 180$$

ACB + (ABC + BAC) – ACB = 180 – ACB (ACB and –ACB cancel each other out)

$$ABC + BAC = \qquad 180 - ACB$$

And

$$ACB + BCD = \qquad 180$$

ACB + BCD – ACB = 180 – ACB (ACB and –ACB cancel each other out)

$$BCD = \qquad 180 - ACB$$

So

$$ABC + BAC = \qquad BCD$$

And of course, an angle of a triangle can't equal 0°. So since ABC and BAC *add up to* BCD, one of them cannot be greater than, or even equal to, BCD.

To generalize, the following holds true for all triangles:

For any vertex of a triangle, its external angle is greater than *either* of the angles of the other vertices. Also, its external angle equals *the sum of* the angles of the other two vertices.

The above are some of the most basic inequalities in geometry, but there are others, too. While it may seem at first that inequalities are harder to deal with than equalities, they do make sense when you examine them carefully. They can even be very useful. There are far more inequalities than there are equalities—at least in mathematics.

Chapter 13: Similar Polygons

Remember congruent triangles? Triangles are congruent if their corresponding sides and angles are the same. There's another concept similar to this, except instead of being the same, the corresponding sides are *proportional*. What does that mean?

Consider the following two polygons:

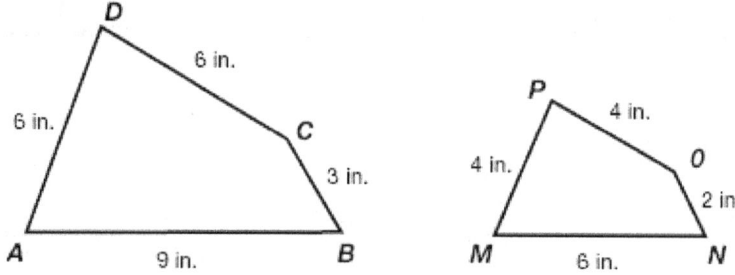

First of all, for two polygons to be similar, their corresponding angles have to be the same size (just like with congruent triangles). Go ahead and assume that's true for ABCD and MNOP above.

Second, the corresponding sides have to be *proportional*. Essentially, this means that you can multiply all the sides of one polygon by *the same number*, and the result will give you all the sides of the other.

Try doing it for ABCD and MNOP as an example. Multiply each side of MNOP by 1.5 and see what happens.

   2 x 1.5 = 3

   4 x 1.5 = 6

   6 x 1.5 = 9

As you can see, multiplying each side of MNOP by the same number—1.5—gives you each *corresponding* side of ABCD. Thus, the two polygons are *similar*.

Of course, just like with congruent triangles, you can move and rotate similar polygons however you want, and they'll still be similar.

So that covers the definition. Now let's talk about some special features of similar polygons—in particular, similar triangles.

Remember how you didn't have to know that *every* corresponding side and angle of congruent triangles were the same to know they were congruent? There's a similar situation for similar triangles. Here are the ways to determine if triangles are similar:

   1. All three corresponding sides are proportional. (SSS or side-side-side)

2. Two corresponding sides are proportional and the angles those sides intersect at are congruent. (SAS or side-angle-side)

3. Two corresponding angles are the same size. (AA or angle-angle)

These allow you to solve problems like the following:

Find the value of x.

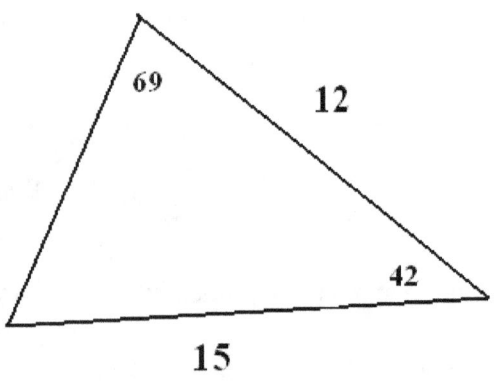 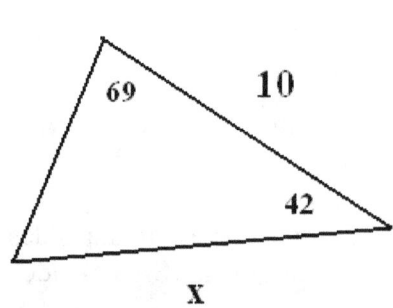

Since two corresponding angles—the ones at the top and the ones on the right—are equal, we know the triangles are similar. (Actually, since all the angles of a triangle add up to 180°, the third angle on each has to be 180 – (69 + 42) = 180 – 111 = 69° for each triangle. Thus, you know that *all* their angles are equal as long as we know that two are.) Therefore, the corresponding sides are proportional. So whatever you multiply 10 by to get 12, you can multiply x by to get 15. Let's call that number *y*. So:

10y = 12

y = 12/10 = 1.2

1.2x = 15

x = 15/1.2 = 12.5

So x is 12.5.

Now, there's one special case of similar triangles you should know about. Consider the following:

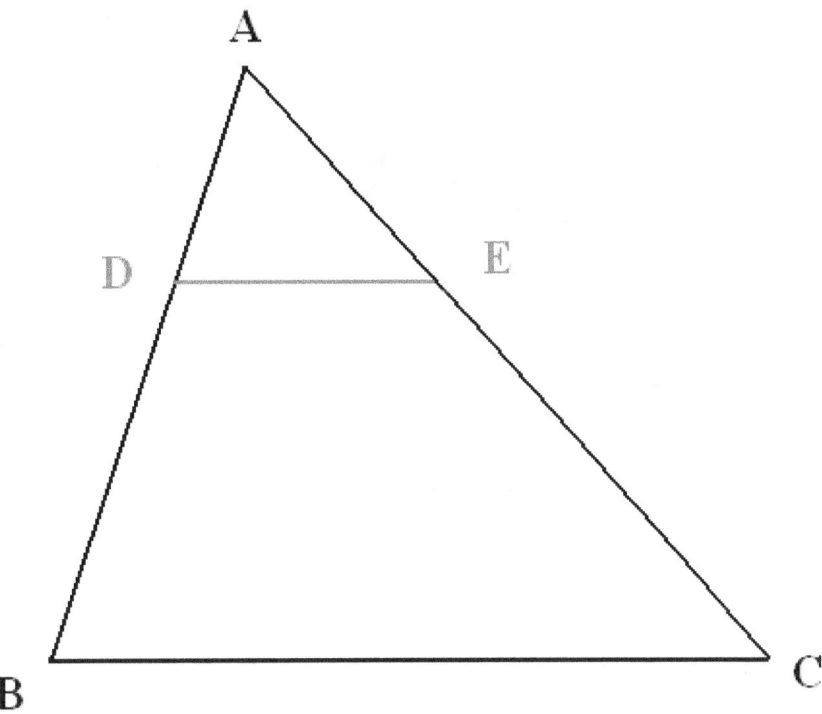

As you can see, the red line DE is parallel to the bottom side BC. When this kind of thing happens, both triangles—ABC and ADE—are similar. To be more general:

> If a line segment is parallel to one of a triangle's sides and intersects the others, the resulting triangle is similar to the original triangle.

As a result of the above, the line segments divide those two sides proportionately. For example, suppose AD was 3 and DB was 6. Then, if EC were 8, you know AE has to be 4
(3 x 2 = 6; 4 x 2 = 8).

Before leaving this section, let's talk a bit about certain shapes that are *always* similar to each other.

First of all, here's a good term to learn: *equilateral triangle*. An equilateral triangle is a triangle where all sides are of the same length, and all angles are the same (This means that all angles are 60 degrees, since they have to add up to 180, and 180/3 = 60).

As you now know, as long as two corresponding angles are the same size, triangles are similar. But for *all* equilateral triangles, *all* the angles are *always* 60 degrees. So every equilateral triangle is similar to every other equilateral triangle.

Here's another special type of triangle: *isosceles triangle*. An isosceles triangle is similar to the isosceles trapezoid we learned about previously. It's a triangle where (at least) two of the sides are equal in length. This also means that the angles opposite the equal sides are equal in size.

Question: Are all isosceles triangles similar? Answer: *No*. One isosceles triangle has angles 50-50-80, while another has angles 30-30-120; no two corresponding angles are equal, so the triangles aren't similar.

Final note: The above rules for determining similar triangles apply *only* to triangles. For instance, although corresponding angles on a rectangle are always equal, all rectangles are *not* similar. On the other hand, all squares *are* similar. Just keep the similarity rules well in mind and you'll be fine.

Chapter 14: Right Triangles

No, you aren't finished with triangles yet!

A *right triangle* is a triangle where one angle is a right angle (90°). Note that it's impossible for a triangle to have two right angles, as 90 + 90 = 180 and then there wouldn't be any degrees left for the third angle.

Here are some examples of right triangles:

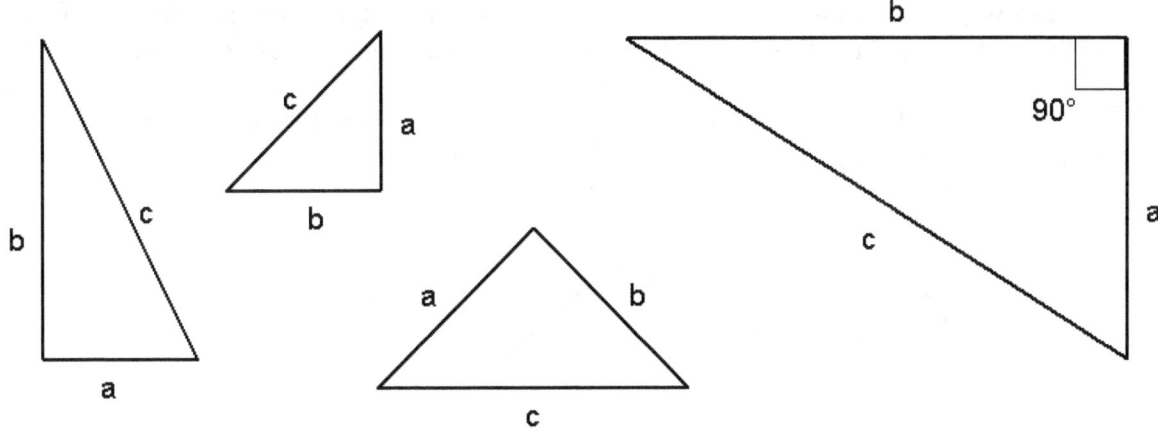

Note that in all the above pictures, the side opposite the right angle is marked "c." By the way, the side opposite a right angle in a triangle is called the *hypotenuse*.

Now, when most people think "right triangle,"they think "Pythagorean Theorem" (if they think anything at all). In case you blocked out all your memories of geometry class, the Pythagorean Theorem states the following:

The square of the hypotenuse is the sum of the squares of the other two sides.

In other words, take any of the right triangles shown above. The value of c^2 is the same as the value of ($a^2 + b^2$). To put it in the way it's usually said:

$$a^2 + b^2 = c^2$$

For example, say that you have a right triangle where the non-hypotenuse sides are of lengths 3 and 4. What's the length of the hypotenuse?

Well, 3^2 is 9, and 4^2 is 16. Add them together and you get 25, whose square root is 5. So the hypotenuse has length 5 ($3^2 + 4^2 = 5^2$).

Of course, not all uses of the Pythagorean Theorem give you a result as pretty as that. For example, what if you're asked to find the length of the hypotenuse of a right triangle when the other two sides have lengths 2 and 3?

$$a^2 + b^2 = c^2$$

$$2^2 + 3^2 = c^2$$

$$4 + 9 = c^2$$

$$13 = c^2$$

$$\sqrt{13} = c$$

Ugly results like that aren't the only issue you can face when trying to use the Pythagorean Theorem. If the sides you have to work with are very long—say 13 or 27—then it might take a long time to square them and add them together. It's a good thing that calculators were invented!

Luckily, there is a way to short-circuit the process, at least for certain types of right triangles. There are two types of these special right triangles. They are classified based on the sizes of the *non*-right angles.

First of all, there are triangles where both non-right angles are 45°. These are actually isosceles right triangles but are sometimes called "45-45-90 triangles." For example:

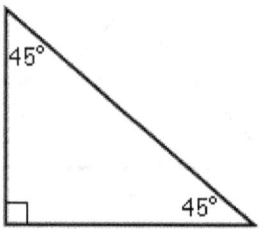

Now, since the two non-right angles are equal, their opposite sides—the non-hypotenuse sides—are also equal. For instance, they could both be size 1. Incidentally, if they were, what size would the hypotenuse be?

The Pythagorean Theorem to the rescue!

$$a^2 + b^2 = c^2$$

$$1^2 + 1^2 = c^2$$

$$1 + 1 = c^2$$

$$2 = c^2$$

$$\sqrt{2} = c$$

Note something about 45-45-90 triangles: since all 45-45-90 triangles have the same angles, they are all similar to each other. So this ratio of sides, $1:1:\sqrt{2}$, holds for every 45-45-90 triangle. Therefore, you can find the side lengths for any 45-45-90 triangle by just dividing the above ratio by the appropriate number, and you can find the hypotenuse by multiplying in the same way. For example, if you have a 45-45-90 triangle where the hypotenuse is $3\sqrt{2}$, you know each side has length 3.

Or, more on point, let's say you have a 45-45-90 triangle where each side is length 182. Taking the square of 182, adding the results together, and then taking the square root of that would be insane. But by using the above ratio, we know what the hypotenuse is: just multiply $\sqrt{2}$ by 182, to get $182\sqrt{2}$.

The other special triangle is a 30-60-90 triangle:

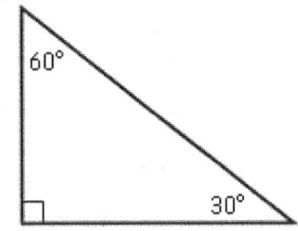

Is there a ratio of sides as there is for the 45-45-90 triangle? Yes, there is—the ratio for a 30-60-90 triangle is $1:\sqrt{3}:2$. To see this, just use the Pythagorean Theorem again:

$$a^2 + b^2 = c^2$$

$$1^2 + (\sqrt{3})^2 = 2^2$$

$$1 + 3 = 4$$

$$4 = 4$$

You can use this ratio the same way you used the 45-45-90 ratio. Let's say you have a 30-60-90 triangle whose hypotenuse is 8. First, you have to find out which number to multiply the ratio by. Since 2 x 4=8, the appropriate number is 4. So the other two sides are lengths: 1 x 4 = 4 and $(\sqrt{3})$ x 4 = $4\sqrt{3}$.

Here's another example: If you have a 30-60-90 triangle with side lengths 500 and $500\sqrt{3}$, using the Pythagorean Theorem can be a bit daunting. But by using the 30-60-90 triangle ratio, just multiply 500 by 2 to get 1000, which is the length of the hypotenuse.

Now, that's…what's that you say? I promised earlier to tell you how to find the length of diagonal line segments in this section? I did, didn't I? Well, like everything else good in life, it's all thanks to the Pythagorean Theorem. For any line segment, such as this one from (0,0) to (2,2):

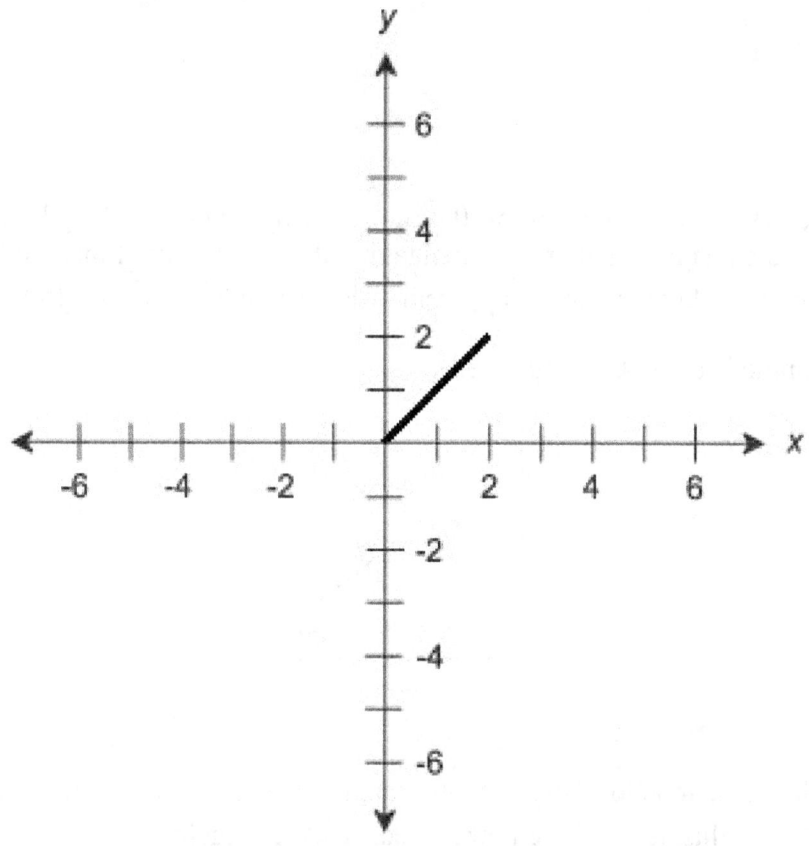

Just draw straight lines and make it the hypotenuse of a right triangle:

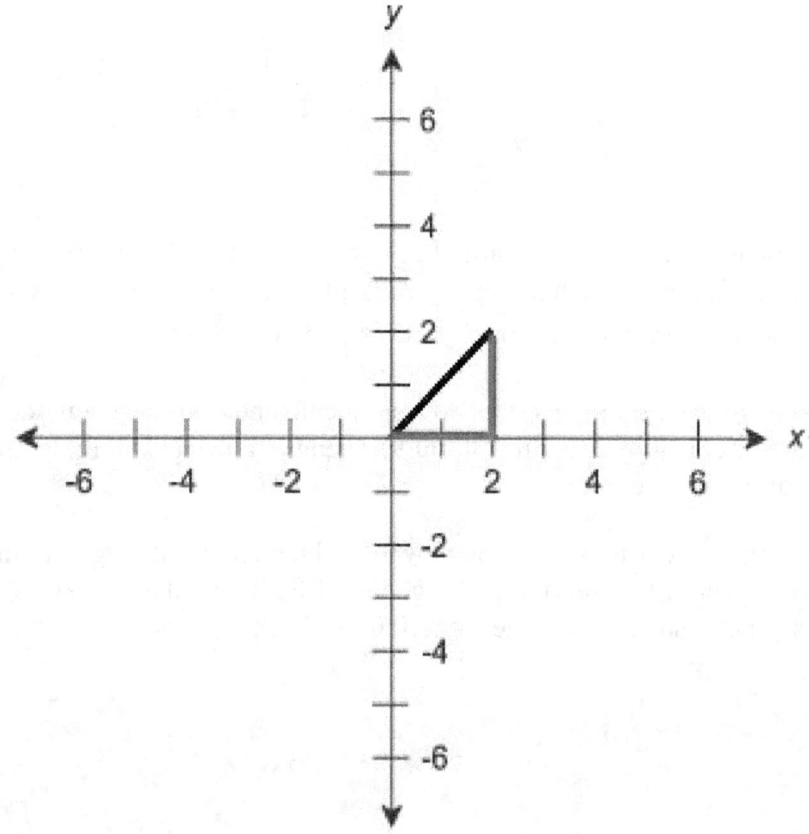

So now you have a right triangle, the non-hypotenuse sides of which go from (0,0) to (2,0) and from (2,0) to (2,2). Since they're straight, you can find their lengths easily: 2. Now just use the Pythagorean Theorem. (L is the length we're trying to find out.)

$$2^2 + 2^2 = L^2$$

$$4 + 4 = L^2$$

$$8 = L^2$$

$$L = \sqrt{8} = \sqrt{4 \times 2} = \sqrt{4} \times \sqrt{2} = 2\sqrt{2}$$

Alternately, of course, you could recognize that since the two non-hypotenuse sides are the same length, you can just utilize the 45-45-90 triangle ratio: multiply $\sqrt{2}$ by 2. Whatever works!

Chapter 15: Circles

You probably know what a circle is already.

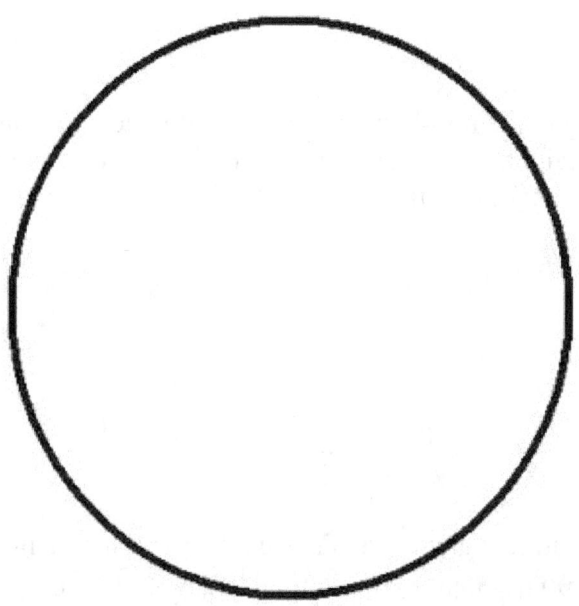

If you want, you can think of a circle as a one-sided shape (Since polygons can't have curved sides, circles aren't polygons).

The technical definition of a circle goes as follows:

> A circle is the shape that consists of every point the same distance (r) from some other point (the circle's center).

An example is generally useful. Consider the point (2,4):

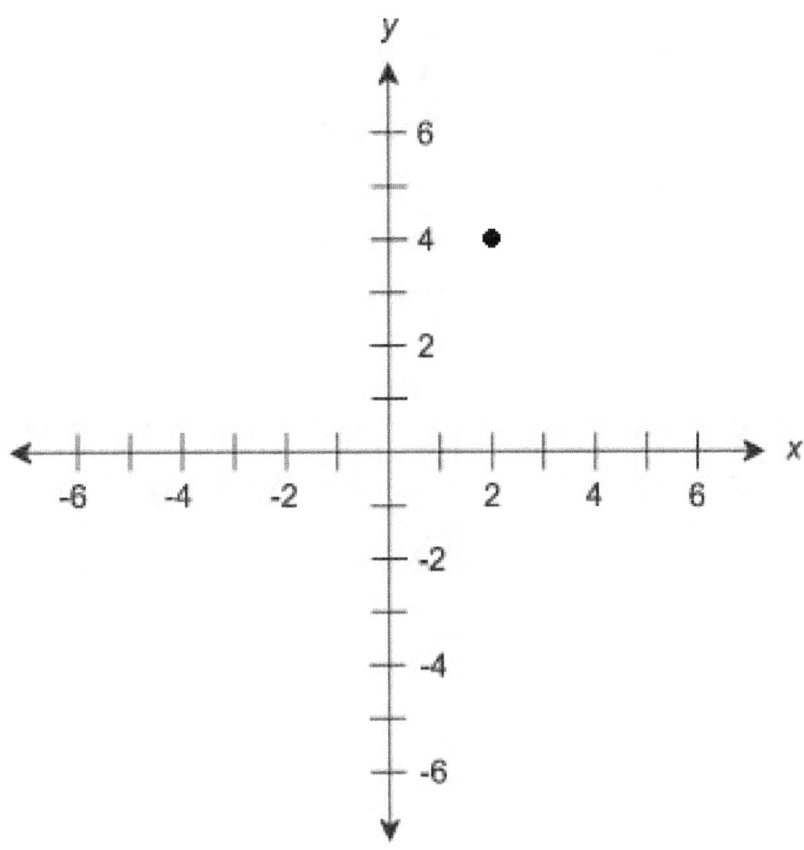

Now, draw a few points that are all 2 away from (2,4)—up, down, left, and right:

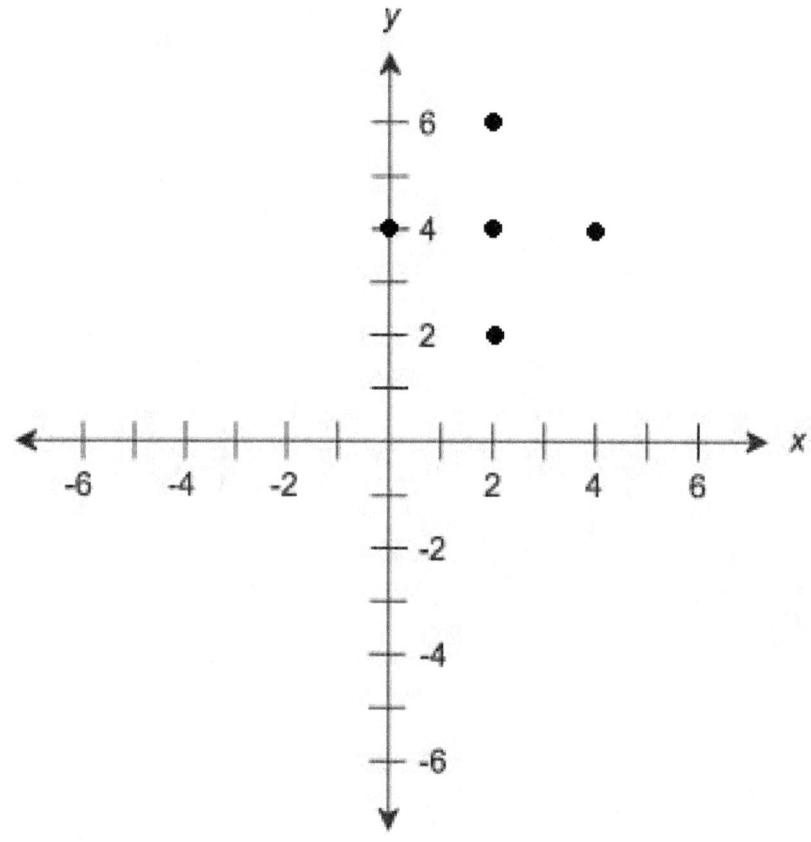

Now you have four of the points that are 2 away from (2,4), but you need the rest too. The easiest way to visualize this is to start by connecting (2,4) to one of the four points we have already, like this:

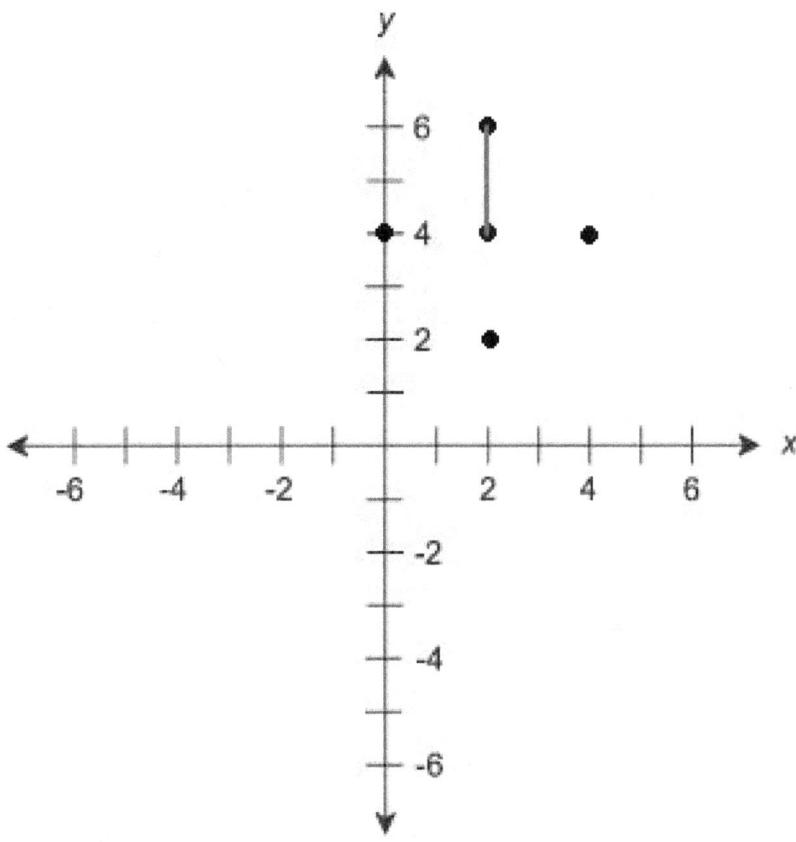

Now imagine that there's a red pen tied to the part of the red line at (2,6). Move the red line to connect to the other three points. Since the red line is of length 2, every point the red pen touches will be 2 away from (2,4). If you imagine the path the red line takes, you will see that the pen will draw a circle in the end:

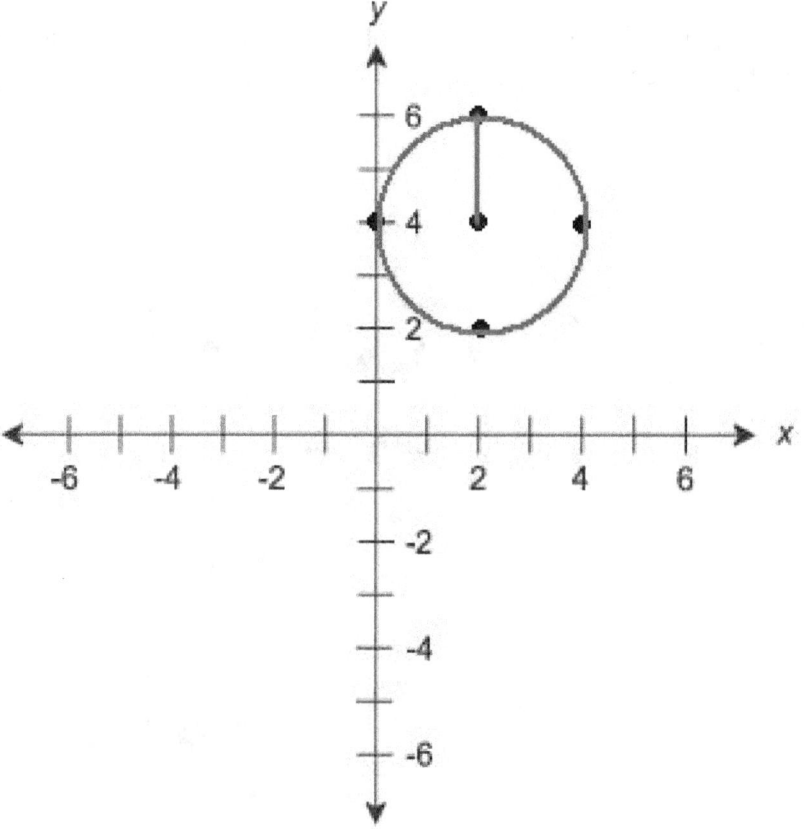

So now you know why the above formal definition works.

Incidentally, since every point on the circle is an equal distance away from the circle's center, every line you draw from the **center** to a **point** on the circle will be the same length. In other words, if you draw a line connecting (2,4) to *any* point on the red circle above, that line will be the same length as the red line—namely, length 2. Any such line is called a *radius*. The straight red line above is an example of a radius.

Here's another important term to know: *diameter*. Any line segment that connects two points on a circle *and passes through the center* is a diameter. For example, the blue line below is a diameter:

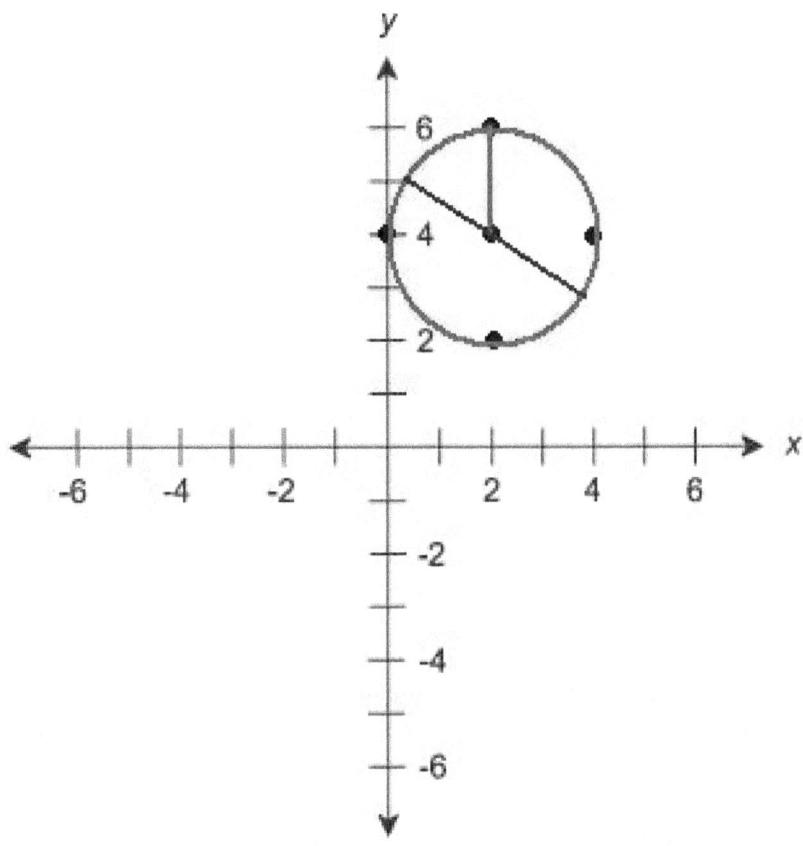

Can you see that any diameter is actually just two radii (the plural of radius) jammed together? Because of that, the length of a circle's diameter is always *twice* the length of that circle's radius.

Of course, it's also possible to connect two points on the circle *without* passing through the center. Any such line is called a *chord*. For example, the green line below is a chord:

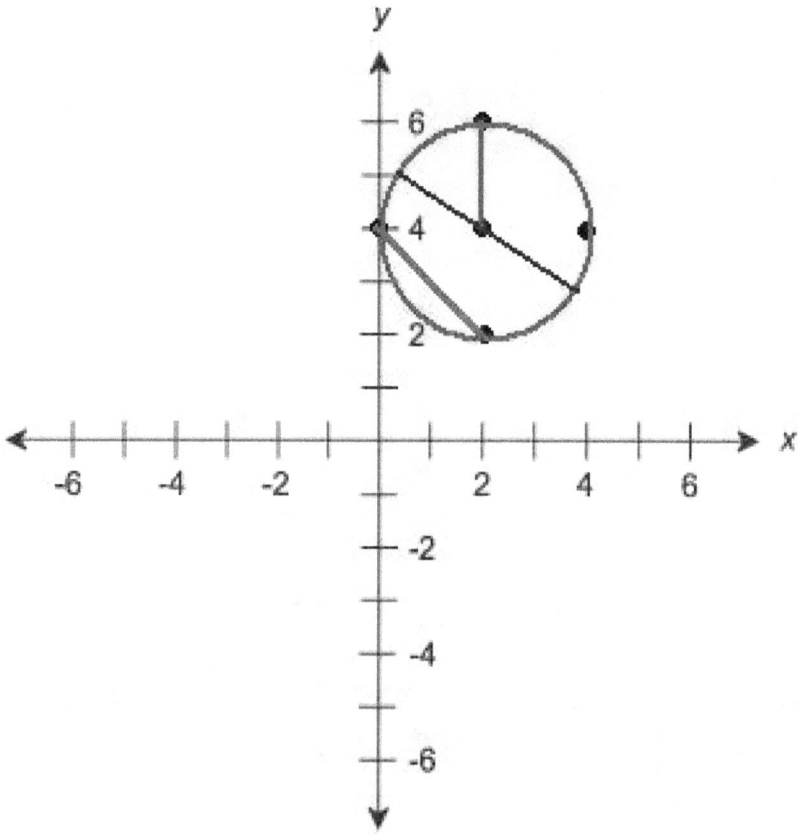

Now let's talk about *circumferences* and *arclengths*.

The circumference of a circle is the distance around the outside. In other words, if you walk along the outside of a circle, the total distance you'll end up traveling is the circle's circumference. So how do you find the circumference of a circle?

The ratio of a circle's circumference to its diameter is actually the same for all circles. In other words, no matter what circle you look at, if you divide its circumference by its diameter, you always get the same number. You might better know this number as *pi*, or π.

That's the good news. The bad news is that π is what mathematicians call an *irrational number*, because it can't be expressed as a fraction. To see why this is, try looking up the exact value of π. You'll get:

3.141592635…

Those dots at the end mean the numbers after the decimal keep on going in the same pattern into infinity. The numbers don't even repeat in a set order—the order keeps changing. They change with no pattern at all. In fact, some random-number generators are based on the digits of π.

So how do you find a circle's circumference? Well, at this point it's almost by definition: since π is what you get when you divide the circumference by the diameter, you get (c is circumference and d is the length of the diameter):

$$\frac{c}{d} = \pi$$

$$c = \pi d$$

In other words, just multiply the circle's diameter by π and you'll get the circumference. Incidentally, since the diameter is just twice the radius r, this can also be expressed as:

$$c = 2\pi r$$

Note that, because π is an irrational number, you can never actually "know" the exact value of a circle's circumference; all you can do is approximate. Most of the time you can just use the symbol.

While the circumference is the distance around the entire circle, the *arclength* is just the distance of one section (or "arc") of a circle. You can calculate the distance of an arclength from the circle's circumference. For example, consider the *arclength* shown in blue.

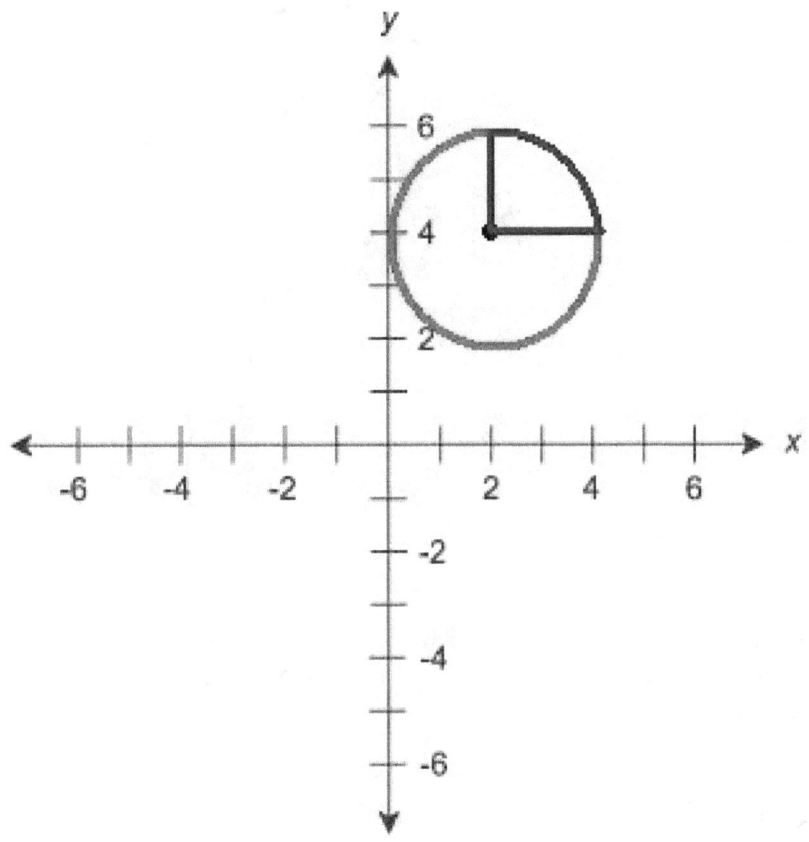

The radius of this circle is 2 (as we know), so its diameter is 4, so its circumference is $\pi d = 4\pi$. As you can see, the arc highlighted in blue is 1/4 of the full circle, so divide the circumference by 4 to get the arclength of the blue arc: π.

In most cases, you have to know the measurement of the angle whose vertex is the center of the circle, also known as the degree measure of the arc. A circle has 360°. Divide the degrees of the arc by 360, then multiply by the circumference to get the arclength.

Chapter 16: Constructions and Loci

There's a particular tool called a *compass*—and no, I'm not talking about that thing that shows you which direction you're going. I'm talking about this:

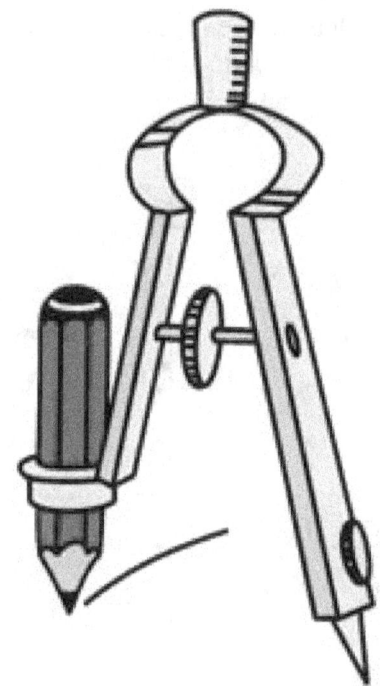

Basically, what a compass lets you do is draw perfect circles—or arcs of circles. You just set it up so that the pencil and the point are a certain distance, and then while planting the point on a piece of paper, move the pencil around.

Now say there are only two tools in the world: a compass and a straightedge (a ruler, the edge of a hardback book, etc.). You can use the straightedge only to draw straight lines, not to measure anything. Just how much can you do with only those tools?

A lot more than you might think. For example, here's how you can use them to draw a line congruent to (the same size as) a given line:

Step 1—Start with a line.

_____

Step 2—Using the straightedge, draw another line somewhere near it. (It doesn't matter how long this line is.)

_____

_____

Step 3—Put the compass's point on the left edge of the first line, and the pencil on the right edge. (In this and future images, the grey line represents the arm of the compass with the point and the blue line represents the arm of the compass with the pencil.)

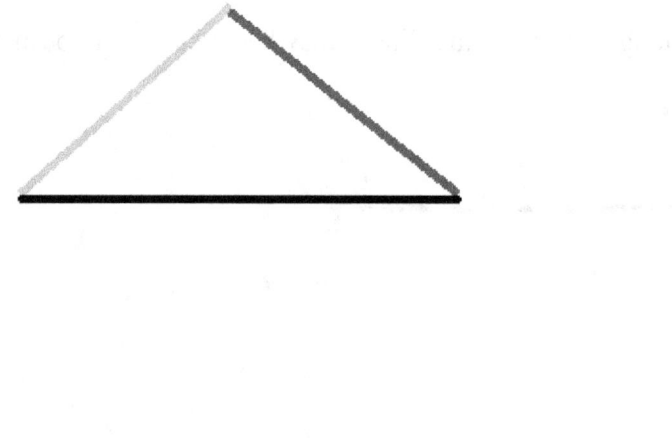

Step 4—Keeping the distance the same, move the compass to the second line. Touch the point to the second line's left edge.

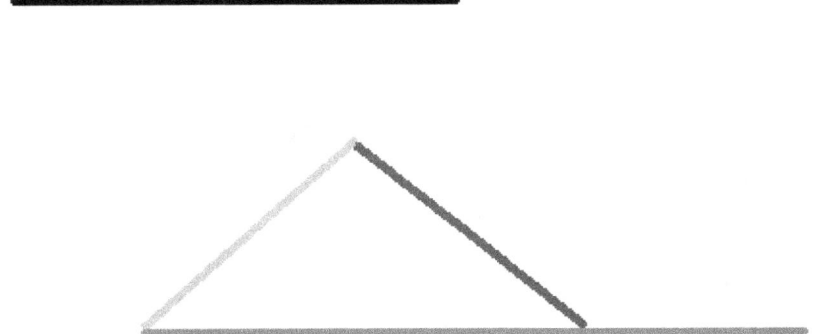

Step 5—Draw a small arc with the pencil. The line segment from the second line's leftmost edge to the point the arc intersects is congruent to the first line.

There are a bunch of other constructions you can do, including drawing a congruent angle and a parallel line, but there isn't enough room to show you those here, so research on your own if you're interested.

Now let's talk about *loci*. (The word is plural—the singular is "locus.")

Let's say I show you a certain point and ask you to find every point that's some fixed distance away from it. If you remember the previous section, you should know that I'm really asking you to find a *circle*. You'd take a compass, put its point on the first point I mentioned, set it to that fixed distance, and then draw a circle. That circle is the *locus* of the problem I gave.

Think of a locus like this: Say you're given a condition (or set of conditions) and asked to find every point that satisfies them. The locus is just precisely those points. In other words,

A *locus* is the set of all points that satisfy (a) given condition(s).

The circle example above is one locus. Here's another:

Problem—Find every point equidistant (the same distance) from the following two points:

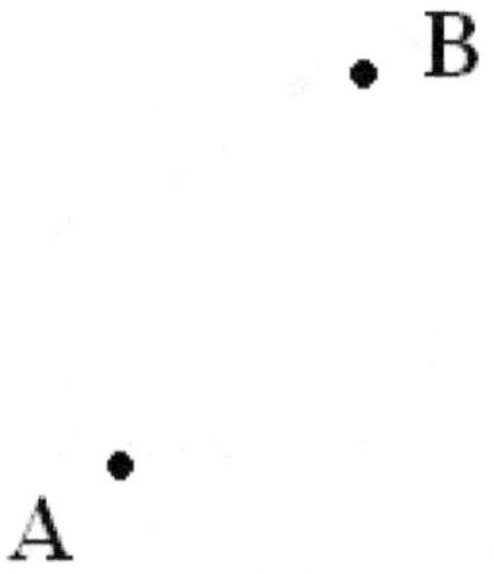

. B

A •

Solution—The basic way to find loci is to find 2 or 3 points that satisfy the given condition(s) and then look for a pattern. In the above example, you can find equidistant points by (1) setting the compass to some arbitrary distance (2) putting its point on point A (3) drawing an arc somewhere between points A and B (4) putting its point on point B and doing the same. The intersection of the two arcs will be a point that's equidistant from A and B.

Do this multiple times and you should get something like this:

Connect those points, and you'll see that the *locus* of this problem is a line:

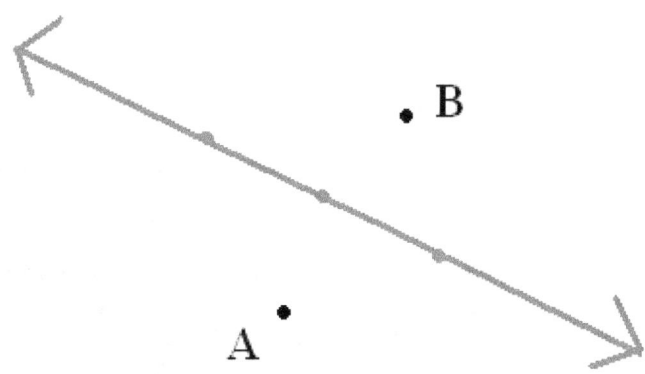

Not just any line, either. Draw a segment connecting A and B, and you'll find that the locus not only divides that segment into two equal parts (in other words, it *bisects* the segment); the line also intercepts the segment at right angles (it's *perpendicular* to the segment). So the locus that's equidistant between two points is the line that bisects the segment connecting those points at right angles.

There are five basic types of loci you should know. You've already learned two—the locus of points at a fixed distance from a single point is a circle, and the locus of points equidistant from two points is a perpendicularly-bisecting line as described above. Here are the other three:

The locus of points equidistant from two intersecting lines is a pair of lines that bisects the *angles* formed by the lines:

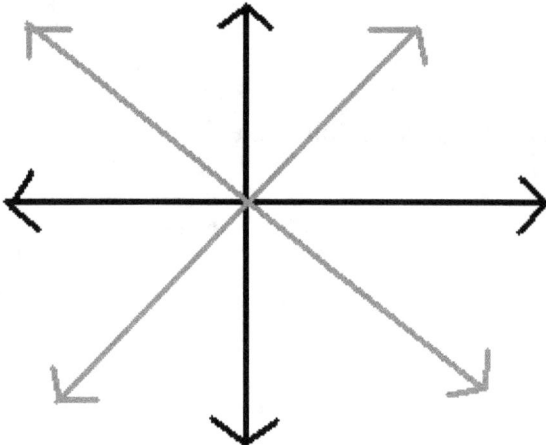

The locus of points equidistant from two parallel lines is one line, parallel to the other two, and halfway between them:

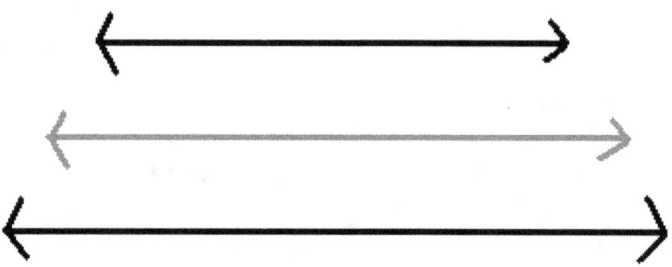

And the locus of points a fixed distance from a single line is two lines, parallel to the given line and the same distance away from it:

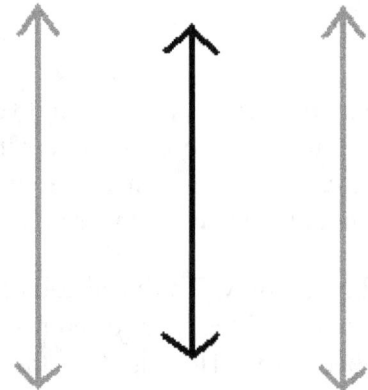

One final note about loci. Sometimes you may be asked to find a locus that fits multiple conditions. In that case, the points have to satisfy *all* the conditions, not just one. To solve those problems, find the loci that fit each individual condition, then find the points common to all those loci. Those points—if any exist—are the locus of the full set of conditions.

Chapter 17: Area of Plane Figures

Let's take a look at a random polygon.

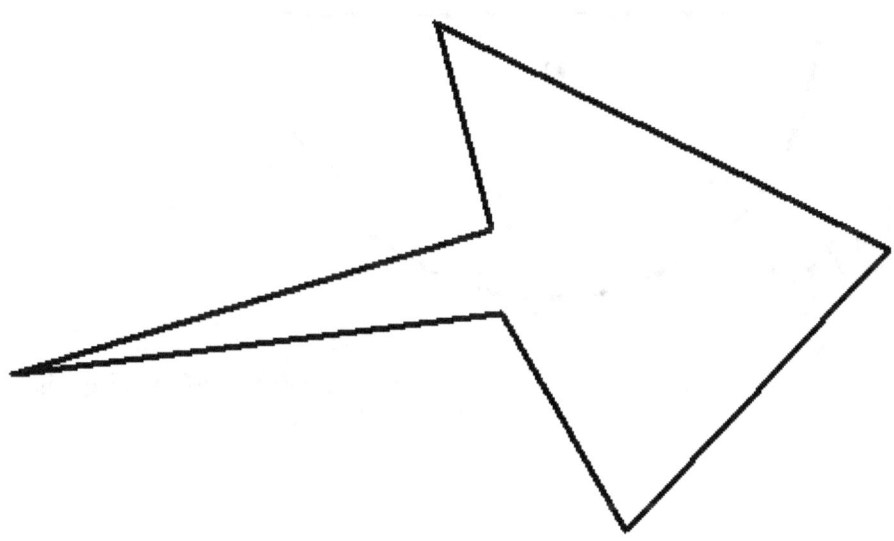

You'll often be asked to find the *perimeter* or the *area* of a polygon. The perimeter of a polygon is distance around it. If you were to walk all the way along a polygon, the perimeter is the total distance you'd travel. To find a polygon's perimeter, just add the lengths of all of its sides together.

The *area* of a polygon is the size of the space *within* the sides. So if you had a room shaped like the polygon above and you wanted to carpet it, the amount of carpet you'd need is the polygon's area. There's no single rule for finding the area of a shape, so in this section, we'll look at all the different kinds of shapes we've learned about so far and see how to find their areas.

Circles

The area of a circle is its radius, squared, times π. Here's the formula:

$$A = \pi r^2$$

For example, if you were asked to find the area of the following circle:

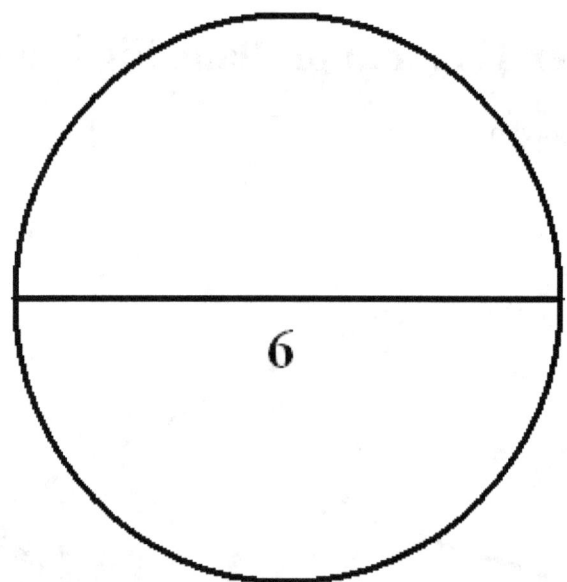

The diameter is length 6. Since the diameter is twice the radius, the radius of the circle is length 6/2 = 3. 3 squared is 9, so the area of the circle is 9π. To put it in equation form:

$$A = \pi r^2$$

$$A = \pi(\frac{d}{2})^2$$

$$A = \pi(\frac{6}{2})^2$$

$$A = \pi 3^2$$

$$A = \pi 9$$

Triangles

The basic formula for finding the area of a triangle goes as follows:

$$A = \frac{1}{2}bh$$

The *b* in the above equation is the length of the triangle's *base*, and the *h* is the triangle's *height*. But what are those?

Let's look at a typical triangle:

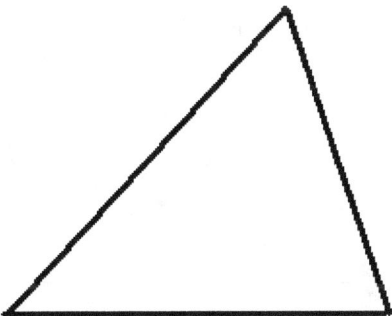

The base is just one of the sides of the triangle. It doesn't matter which side you pick, so let's just pick the bottom side.

Once you have a base picked out, you need to draw a line from the opposite vertex to the base—and that line has to intersect the base at a right angle (that is, they have to be *perpendicular*). That line will be the *height*. In the above example, you get:

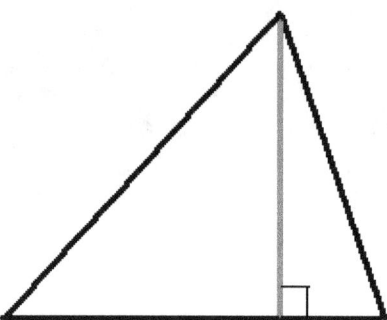

So in the above picture, if the bottom side is the *base*, the red line is the *height*. So to find the area, just multiply the length of the base by the length of the height, then divide by 2.

Let's look at another example. Suppose you were asked to find the area of the following triangle:

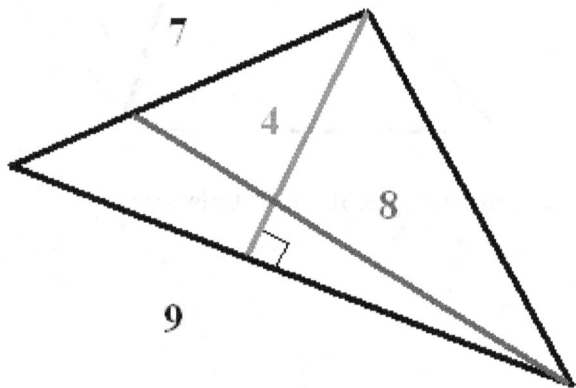

First, figure out which line to use as the height. The blue line cannot be the height, because it doesn't intersect the opposite side at a right angle. The red line, however, does. So the height is the red line, and the base is the side the red line intersects. So:

$$A = \frac{1}{2}bh$$

$$A = (\frac{1}{2})9 \times 4$$

$$A = (\frac{1}{2})36$$

$$A = 18$$

So the area of the triangle is 18.

Parallelograms

You can find the area of all parallelograms the same way. Take a look at a parallelogram:

Now, draw one of its diagonals:

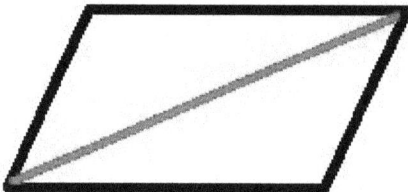

See how a parallelogram is really just two congruent triangles put up against one another? That means that the area of a parallelogram is just twice the area of one of those triangles, or:

$$A = bh$$

b and *h* are the base and height of the parallelogram, same as with a triangle.

Let's look at an example. Find the area of the following parallelogram:

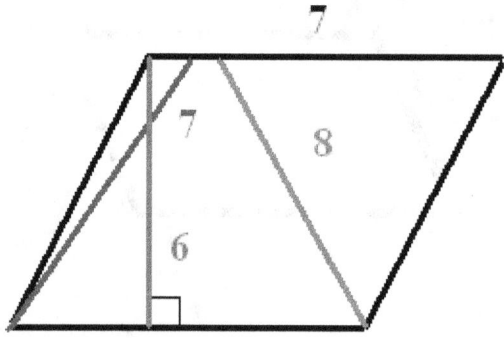

This time, only the green line intersects a side at a right angle. Since opposite sides of a parallelogram are equal in length, the side the green line intersects must also be length 7. Therefore, the area of the parallelogram is:

$$A = bh$$

$$A = 7 \times 6$$

$$A = 42$$

For the special types of parallelograms, calculating the area becomes simpler. Since all four sides of a rhombus are equal, the area is just any one side times the height:

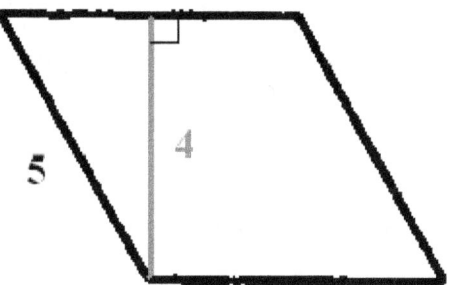

$$A = sh$$

$$A = 5 \times 4$$

$$A = 20$$

As for rectangles, because the sides of a rectangle already intersect each other at right angles, you just have to multiply one side by another side that's *not* parallel to it (that is, multiply its *length* by its *width*):

$$A = lw$$

$$A = 11 \times 4$$

$$A = 44$$

And since a square has the special properties of both rhombuses and rectangles, all you have to do is square (yes, this is a geometry pun) the length of one side:

9

$$A = s^2$$

$$A = 9^2$$

$$A = 81$$

Trapezoid

Trapezoids are a bit more complicated. The area of a trapezoid is given by the following formula:

$$A = \frac{a + b}{2}h$$

h, same as always, refers to the trapezoid's height. *a* and *b* here are the lengths of each of the trapezoid's *parallel* sides. So you are averaging the length of the two parallel sides and multiplying it by the height.

For example, find the area of the following trapezoid:

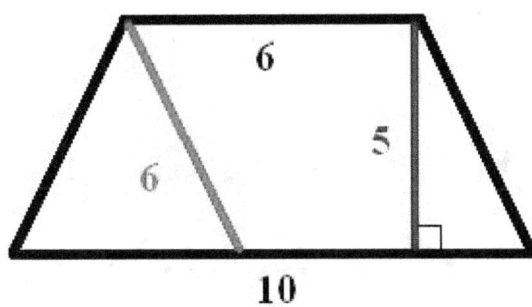

Again, the blue line is the height because it intersects the base at a right angle. So:

$$A = \frac{a+b}{2}h$$

$$A = \frac{6+10}{2}5$$

$$A = \frac{16}{2}5$$

$$A = 8 \times 5$$

$$A = 40$$

Kites

The easiest way to calculate the area of a kite does not involve its sides. It actually involves the kite's diagonals. The formula is:

$$A = \frac{p \times q}{2}$$

Where p and q are the lengths of each of the kite's diagonals.

For example, find the area of the following kite:

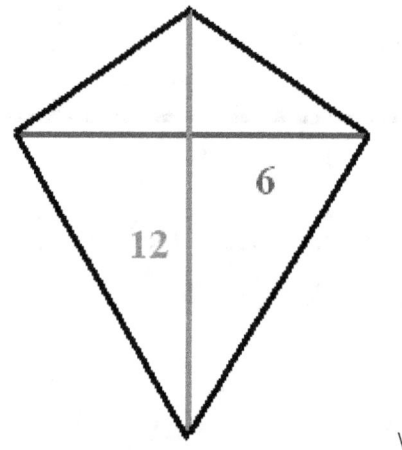

$$A = \frac{p \times q}{2}$$

$$A = \frac{12 \times 6}{2}$$

$$A = \frac{72}{2}$$

$$A = 36$$

That's a lot of formulas. But that's just the way math works sometimes. Do your best. You'll be fine.

Chapter 18: Areas and Volumes

You're almost at the end! There's only one more topic to cover here, and that is the dreaded…*three-dimensional objects*! (Okay, maybe not so dreaded.)

Throughout these sections, every object you've learned about so far has been a two-dimensional object, with only length and width. But the world we live in has *three* dimensions—length, width, and height. Geometry deals with three-dimensional objects too. In this section, you'll go over them one at a time. First, consider *surface area* and *volume*.

The surface area of a three-dimensional object is just what it sounds like—the area of the surface of the object. So if you're holding a ball, for instance, the total size of the space you can touch is the surface area. The *volume*, on the other hand, is the space *inside* the object; for example, the total amount of liquid a given ball could store (assuming its surface is infinitely thin) is the sphere's volume.

Put another way, *surface area* is a little like the wrapping paper on a present. *Volume* is everything inside the wrapping paper—present, box, air, and all.

Note that a three-dimensional object's *surface area* is kind of like a two-dimensional object's *perimeter*, while the volume is kind of like the area.

Now, with that out of the way, let's move on to three-dimensional objects!

Rectangular Prisms

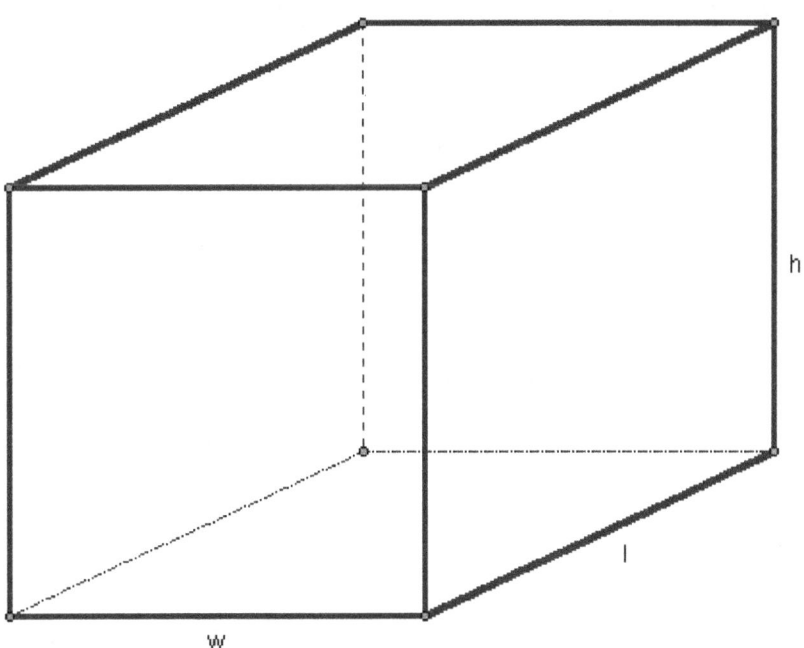

As you can see, a rectangular prism is basically six rectangles glued together to make a single three-dimensional object. A *cube* is a special kind of rectangular prism, one that has six *squares* glued together; alternately, it's one where the width, length, and height (represented by *w*, *l*, and *h* above) are all the same.

You can see that finding the surface area is just a matter of adding up the areas of the six rectangles that make up the rectangular prism. Because each opposite side of the prism has the same area, you can simplify this to:

$$A = 2wl + 2wh + 2lh$$

You find the volume by multiplying the width, length, and height together. In equation form, it looks like this:

$$V = wlh$$

For a cube, since the width, length, and height are all the same, the equations are simpler (*s* refers to any side of the cube):

$$V = s^3$$

Cylinders

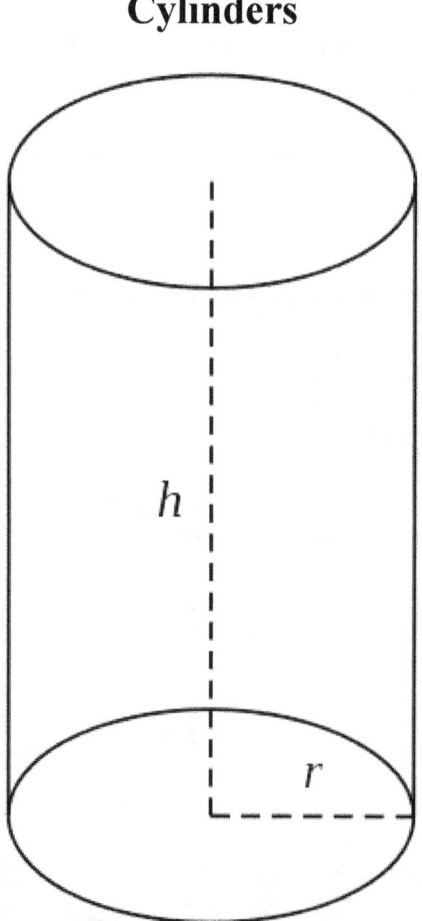

As you can see from the above image, a cylinder is two congruent circles (circles with the same radius) connected by a rounded prism. Think of it as a can of soup.

The surface area of a cylinder is the area of the circle on the top plus the area of the circle on the bottom, plus the area of the prism connecting them. The areas of the circles, as you know, are each πr^2. The area of the prism is $2\pi rh$. h is the height of the cylinder, as marked above. $2\pi r$ is the circumference of the circle. So this is the formula for the surface area of a cylinder:

$$A = 2\pi r^2 + 2\pi rh$$

The volume is simpler: take πr^2 (the area of one circle) and multiply by h (the height of the cylinder). In equation form:

$$V = \pi r^2 h$$

Pyramids

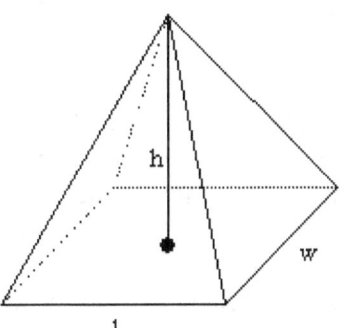

You know the Great Pyramids in Egypt, right? There are actually a large variety of pyramids, depending on what the *base*—the shape at the bottom of the pyramid—is. Other important pyramid terms are the *apex* (the point at which all the sides meet) and the *height* (the distance from the base to the apex).

The surface area of a pyramid depends on what the base is. In general, you have to add the area of the base to the area of all the triangles that make up the pyramid's sides. If those triangles are all congruent, though, then you get:

$$A = B + \frac{1}{2}PL$$

B is the area of the base, P is the *perimeter* of the base, and L is the height of one of the side triangles. The volume is less involved (h is the pyramid's height):

$$V = \frac{1}{3}Bh$$

If the pyramid is a square pyramid—its base is a square—the formulas become more sensible (s is the length of one side of the square base):

$$A = s^2 + 2sL$$

$$V = \frac{1}{3}s^2h$$

Cones

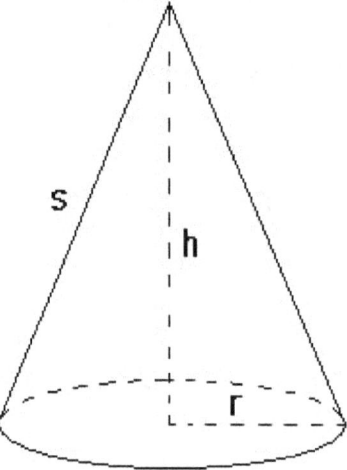

A cone is kind of a like a pyramid with a circle at its base. (It's also the shape of a dunce hat or waffle cone that unfortunately has no ice cream in it.) In the picture above, r is the radius of the base circle, h is the distance from the base to the apex at the top, and s is the distance from one of the points on the outside of the base circle to the apex.

For the surface area, add the area of the base circle (πr^2) to the area of the rest of the cone ($\pi r s$):

$$A = \pi r s + \pi r^2$$

To get the volume, multiply πr^2 by the height and then divide by 3:

$$V = \frac{1}{3}\pi r^2 h$$

Spheres

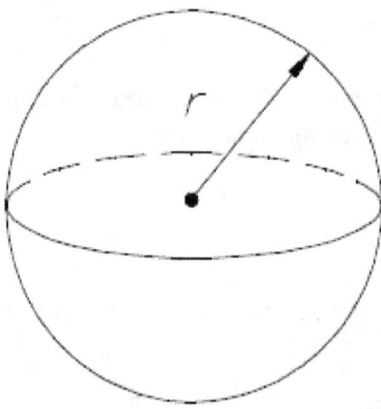

Finally, to wrap up, let's talk about spheres. You can think of a sphere as every point a fixed distance r away from a single point in three-dimensional space, much like a circle is every point a fixed distance r away from a single point in *two*-dimensional space.

To find the surface area and volume of a sphere, all you have to know is that distance r. Here are the formulas:

$$A = 4\pi r^2$$

$$V = \frac{4}{3}\pi r^3$$

Important note: While you square r in the surface area formula, you *cube r* in the volume formula.

Again, I know that's a lot of formulas, but that's life (or mathematics). Of course, you can always come back to this page if you forget any of them.

Conclusion

Twelve sections and much geometrical knowledge later, what have you learned?

There are the individual definitions and formulas, of course. Those are very important. But more than any one term or area formula, geometry is really about *evidence*. This short course has tried as much as possible to show why the truths of geometry are the way they are and how they relate to each other. While geometry may seem strange and obscure, everything in it does actually make sense, and everything has a reason for being the way it is. Geometry is more of a way of thinking than a set of stuff you have to memorize. Let's hope that you've seen at least a little of that here.

About Minute Help

Minute Help Press is building a library of books for people with only minutes to spare. Follow @minutehelp on Twitter to receive the latest information about free and paid publications from Minute Help Press, or visit minutehelpguides.com.